T0343880

Project Management for Drug Developers

Project managers in drug development are the driving force behind the coordination of efforts. This book provides a practical reference for project managers in the pharmaceutical and biotech drug development industry, with the goal of assisting in creating an efficient and effective team structure and environment. The text details the role of project managers at each stage of drug development, the key interfaces that the project manager will need to work closely with, and essential tools of the trade, including frequently used techniques and methodologies. This book is useful for both entry-level and advanced-level project managers, as well as non-project managers from other functions.

Features

- Includes authors' recent experience with improved tactics and technologies/software at various stages of drug development.
- Provides the most up-to-date and best practices, techniques, and methodologies in project management.
- Details the role of the project manager at each stage of drug development, including working with the key interfaces throughout the process.
- Diverse audience, including non-project managers in clinical development, clinical operations, regulatory affairs, medical affairs, clinical pharmacology, and biostatistics.
- Provides templates and timelines for critical paths from development to commercialization and has potential as a textbook on relevant courses.

Drugs and the Pharmaceutical Sciences

A Series of Textbooks and Monographs

Series Editor

Anthony J. Hickey

RTI International, Research Triangle Park, USA

The Drugs and Pharmaceutical Sciences series is designed to enable the pharmaceutical scientist to stay abreast of the changing trends, advances, and innovations associated with therapeutic drugs and that area of expertise and interest that has come to be known as the pharmaceutical sciences. The body of knowledge that those working in the pharmaceutical environment have to work with, and master, has been, and continues, to expand at a rapid pace as new scientific approaches, technologies, instrumentations, clinical advances, economic factors, and social needs arise and influence the discovery, development, manufacture, commercialization, and clinical use of new agents and devices.

High Throughput Screening in Drug Discovery
Amancio Carnero

Generic Drug Product Development: International Regulatory Requirements for Bioequivalence, Second Edition
Isadore Kanfer and Leon Shargel

Aqueous Polymeric Coatings for Pharmaceutical Dosage Forms, Fourth Edition
Linda A. Felton

Good Design Practices for GMP Pharmaceutical Facilities, Second Edition
Terry Jacobs and Andrew A. Signore

Handbook of Bioequivalence Testing, Second Edition
Sarfaraz K. Niazi

FDA Good Laboratory Practice Requirements, First Edition
Graham Bunn

Continuous Pharmaceutical Processing and Process Analytical Technology
Ajit Narang and Atul Dubey

Project Management for Drug Developers
Joseph P. Stalder

For more information about this series, please visit: www.crcpress.com/Drugs-and-the-Pharmaceutical-Sciences/book-series/IHCDRUPHASCI

Project Management for Drug Developers

Edited by
Joseph P. Stalder
Founder, Groundswell Pharma Consulting
San Diego, USA

CRC Press is an imprint of the
Taylor & Francis Group, an informa business

First edition published 2023
by CRC Press
6000 Broken Sound Parkway NW, Suite 300, Boca Raton, FL 33487-2742

and by CRC Press
4 Park Square, Milton Park, Abingdon, Oxon, OX14 4RN

ISBN: 978-1-032-12668-5 (hbk)
ISBN: 978-1-032-12920-4 (pbk)
ISBN: 978-1-003-22685-7 (ebk)

DOI: 10.1201/9781003226857

Typeset in Times
by codeMantra

Contents

Foreword

Development of pharmaceutical products is one of the most complex and expensive challenges for teams to undertake. It requires the identification of a novel approach to treat a sickness or disease, executing a plan for confirming compound utility, manufacturing of trial drug lots (and subsequent scale-up), non-clinical testing to support safety studies, determining the effectivity of the drug in animal models, and preparing information for submission to regulatory agencies to allow the work to continue. Years of evaluation are needed before a company can even request the initiation of a small clinical trial. Challenging regulatory requirements must be met, and considerations of those for each target geographic region need to be addressed. Competitive pressures for speed to market drive taking the most efficient pathway possible. It is a daunting task for even the most experienced companies.

Project management methodologies are being used in all functional areas of the development process to meet these challenges. Project managers are assigned to drug asset development, CMC activities, clinical trials, and regulatory submission preparations. Project management for the entire product development program requires alignment of the activities of all the functional teams to keep the program on track. This herculean task requires using the best approaches and continually making improvements.

There are some challenges that make pharma product development more difficult than other industries. The large number of cross-functional teams needed at every stage of development can lead to communication challenges and unnecessary delays. Since healthcare solutions are global, product development is also global with teams spread across time zones, supply chains, and cultural working norms. Making management even more difficult, many functional activities are outsourced to specialty firms, which can lead to issues and delays. The testing of the target molecule(s) for effectivity and safety while development is ongoing can require pivoting when unexpected results occur. Finally, regulatory requirements differ in each country and are continually being upgraded, leading to new challenges throughout the development cycle to ensure a smooth pathway to regulatory approvals. Each of these complexities requires particular attention and are the bedrock challenges that project management addresses.

This book discusses the challenges that project managers face, and approaches and solutions to improve project management results. It provides a framework for the overall processes that must be managed and details on how workstreams interact. Beginning with descriptions of how project management overlays on the drug development process, specific topics include different roles for project managers, managing teams, using a project management office approach to improve project management and dealing with project portfolios. Then, information on several best practice approaches for handling major challenges is detailed. The authors tackle topics including management of international projects, information systems, creation of key plans, and successful regulatory submissions.

The third section of this book provides a new way of thinking about the key elements that drive managing projects. Termed as GRIDALL, it provides a framework to link these key elements: goals, risks, issues, decisions, actions, and lessons learned. Understanding how to deal with these elements is integral for a project manager to successfully support the development of a pharma product. The framework shows how setting of program goals and alignment of all stakeholder goals is necessary to achieve program milestones. Since product development is handled by multiple functional teams (both internal and external), ensuring the functional and individual goals linked to program goals is needed to efficiently complete the necessary work.

Integrated scheduling by project management is the means by which sub-goal alignment can be done. The alignment helps determine prioritization and timing, especially for interdependent activities and deliverables. Risk planning is performed to reduce the likelihood that unexpected events or missed deliverable timelines have significant impacts on the interconnected activities or lead to delays. Often the cascading effect of one, seemingly small, deliverable change (timing, quality, or result) has a drastic impact on the overall program. Fully integrated program management reduces the likelihood of this occurring by keeping stakeholders aligned on progress, potential risks, realized issues, and any cross-functional impacts. Effective decisions and appropriate actions can therefore be taken at the right time.

Communication is the underlying mechanism for alignment and synchronization of product development teams. It is also the most difficult part of leading a pharma program. Each stakeholder group and functional team requires filtered information that is both current and accurate. They also need help to identify how changing conditions impact their workstreams, as well as how their team's progress impacts other workstreams.

Successfully dealing with the unpredictability of complex drug development can only happen when the key elements of the parallel workstreams are monitored, communicated effectively, and then addressed efficiently. The GRIDALL approach provides a roadmap for how to successfully manage these challenges and approach the cross-functional communications that are so critical.

This outstanding book is the result of collective insights from pharmaceutical product development project management experts who share their experiences and offer practical approaches to meet the challenges of drug development. It provides a comprehensive guide for project managers that are new to biopharma product development and a sound reference for experienced practitioners.

Scott D. Babler MA, MBA, CSSBB, PMP
Principal Consultant, Integrated Project Management Company, Inc.
Editor and Co-Author of: Pharmaceutical and Biomedical Project
Management in a Changing Global Environment (2010)
John Wiley & Sons Inc.

Preface

In the face of relentless competition and evermore-demanding expectations from regulators, patients, prescribers, and payers, the biopharmaceutical industry constantly strives for better, faster, and cheaper product development. But drug development is complex and risky, and the inevitable challenges that arise during the multi-year lab-to-launch process can only be overcome through careful coordination, collaboration, and communication. Hence the rise of the project management discipline in the industry over the last 30 years. Companies have realized the value that Project Management adds by converting complex strategies into actionable plans, and efficiently executing those plans across cross-functional boundaries. Project Management addresses several critical success factors for pharmaceutical product delivery (e.g., collaboration of multiple disciplines, coordination of business processes, communication across multiple stakeholders), and the unique combination of knowledge, skills, and behaviors that a Project Manager provides has become a crucial part of any drug development project.

There are several great books available about project management in the biopharmaceutical industry, but none focus specifically on the role of a Development Project Manager – the ones who manage assets from first-in-human studies to submission of a market application. This book aims to address this audience, including both experienced drug developers who are looking to transition into project management and practicing PMs who are looking for best practices in managing clinical-stage assets.

We aimed for this book to be very practical, with modern tools, techniques, templates, and methodologies that are applicable to situations a Development PM is likely to encounter throughout the stages of drug development. In fact, when authoring the book, we set a standard of 20% theory and 80% practical – just enough theory to make sense of the practice. We want the reader to be able to read a chapter and then apply what he/she learned the next day they go to the office.

We realize that no company is the same, and the expectations for each PM may differ slightly from company to company, and even from project to project. Nonetheless, drawing on experiences from over 30 biopharma companies – large and small, startup and mature, regional and international, small molecule, biologics, devices – we were able to agree on what we think are "best practices" that apply to most PMs. The output is a collection of examples and recommendations on how to address common challenges and capitalize on potential opportunities, regardless of the company or the project.

One caveat to the reader is that nomenclature is notoriously inconsistent between companies. What we may call a Core Team in this book may be called something else at your company. When we refer to the Safety department, it may be called Drug Safety Evaluation, Preclinical Operations, Toxicology, or some other variation. You may have to translate from our description to the term that is used in your company.

This book is organized into three parts. Part 1 sets a foundation of the drug development process and the role Project Managers play in drug development. Knowing that PMs work within the context of a project team and organizational management constructs, we highlight the interfaces that PMs have with team members, key business stakeholders, and governing bodies. A deep understanding of the drug development process and working ability to effectively manage these interfaces not only builds a PM's trust and credibility with the team, but also makes the PM's job easier because there will be fewer executional surprises to deal with.

Part 2 of this book dives deep into contemporary topics in drug development. These topics are being actively discussed within companies, at conferences, and in published articles, and there is no consensus on how they should be treated, nor will there ever be. What we present here is a synthesis of experiences into an aligned recommendation for how to address the issue in biopharma.

Part 3 of this book describes GRIDALL, a comprehensive framework for managing projects. After introducing the concept of GRIDALL and how it can be applied to biopharma, each chapter dives deep into the theory and practice of setting goals, managing risks, addressing issues, making decisions, taking action, and capturing lessons learned. This chapter has many tools and templates built in that can serve as a starting point for building a project management function in an organization or as an assessment tool to determine the completeness of an existing project management function.

If you've purchased this book, you already know the value that drug development companies bring to society in the form of improved health, longer lives, and happier living. What this book adds is that the consistent application of project management best practices will expedite the development of these important drugs so they can reach patients sooner. We hope that you will continue to carry the torch of progress in the field of project management so that our craft becomes more finely tuned and effective at delivering new medicines to the world.

Glossary of Terms and Abbreviations

Activity: A distinct, scheduled portion of work performed during the course of a project. An activity is an action statement with a start date and duration. When an activity is assigned to someone, it becomes a *task*.

ADME: Absorption, Distribution, Metabolism, and Excretion.

Analogous Estimating: Estimating using similar projects or activities as a basis for determining the effort, cost, or duration of a current one. Usually used in top-down estimating.

API: Active Pharmaceutical Ingredient.

Asset: A therapeutic entity (e.g., small molecule, biologic, cellular, or genetic therapy) that is being developed into a marketed drug.

Assumption: Something taken as true without proof. In planning, assumptions regarding staffing, complexity, learning, curves, and many other factors are made to create plan scenarios. These provide the basis for estimating. Remember, assumptions are not facts. Make alternative assumptions to get a sense of what might happen in your project.

Asset Development Plan (ADP): Sometimes called an Integrated Development Plan (IDP), Integrated Product Plan (IPP), or Product Development Plan (PDP); the ADP is the plan for the overall development of a product. It includes the medical need, target profile, business case, technical strategy, regulatory strategy, and commercial strategy.

Asset Development Team: A group of individuals brought together to maximize the value of an asset.

Bioavailability: The rate and extent to which a drug is absorbed or is otherwise available in the body.

Bioequivalence: Scientific basis on which two drug products are determined to be interchangeable. To be considered bioequivalent, the bioavailability of two products must not differ significantly when the two products are given in studies at the same dosage under similar conditions.

Biologics License Application (BLA): In the US, a market application submission equivalent to a New Drug Application for biologics. BLAs are reviewed by CBER within the FDA.

Business Case: The information that describes the justification for the project. The project is justified if the expected benefits outweigh estimated costs and risks. The business case is often complex and may require financial analysis, technical analysis, organization impact analysis, and a feasibility study.

Clinical Development Plan (CDP): A document that defines the preferred strategy for clinical drug development to enter and capture a defined market. It specifies the evidence that should be collected and targets a development time from entry into clinical development to submission of a market application.

Clinical Candidate: Also known as Lead Candidate. A new chemical entity selected for evaluation in the clinic for the treatment of a disease.

Chemistry, Manufacturing, and Controls (CMC): 1. In the context of a sponsor company, the department that oversees the development, manufacture, and analytical testing of drug substance and drug product. Also known as Technical Operations or Pharmaceutical Research & Development (PR&D). 2. A section in a market application that describes the composition, manufacturing process, and control of the drug substance and drug product.

Cost of Goods Sold (COGS): Refers to the cost of manufacturing the final drug product. Typically described in terms of cost per unit (tablet, capsule, or vial).

Clinical Study Report (CSR): The document that describes the methods, results, and interpretation of a clinical trial.

Common Technical Document (CTD): A common format for all submission dossiers to regulatory authorities worldwide following agreed ICH standards.

Continuous Improvement: An approach to capturing lessons learned by reviewing activities as they occur. Also known as kaizen. Compare to *post-mortem.*

Contract Development and Manufacturing Organization (CDMO): An organization that is contracted by a sponsor to execute the design, development, and/or manufacture of a new chemical entity.

Contingency Plan: A set of activities intended to be used in the event that a specific trigger occurs (i.e., a risk materializes into an issue). Compare with *Mitigation* Plan.

Contract Research Organization (CRO): A company that is contracted by a sponsor to execute specific pieces of the drug development process, particularly the execution of clinical trials (e.g., a study gets outsourced to a CRO).

Critical Path: The sequence of dependent activities in a project that determines the duration of a project. Critical path tasks have zero slack time. If any activity on the critical path is delayed, the overall project duration will extend.

CTMS: Clinical Trial Management System.

Data Cut Off (DCO): The last day for data to be included in a periodic or aggregate safety reports or, in the case of oncology studies, for efficacy analyses.

Decision: A commitment made by an organization to take a certain action

Decision-Making Framework: A conceptual guide that describes the governance structure, remits, and roles for each governing body in an organization

Dependency: A relation between project activities such that one requires input from the other. The relationship can exist within a single project or between projects.

ERP: Enterprise Resource Planning system.

Enterprise Software: A software application used by a large portion of an organization. For example, a project management solution becomes enterprise software when it includes timesheets that all employees are expected to complete.

Escalation: The process of notifying senior management of critical risks and issues so that an appropriate response/decision can be implemented.

Food and Drug Administration (FDA): The US federal agency responsible for enforcing the Federal Food, Drug, and Cosmetic Laws and Regulations.

FTE: Full-Time Equivalent.

Goal: The generally desired outcome or output for a project. Goals reflect directly on the key objectives of the project.

Governance: The process, decision rights, and accountability at various levels of the institution to manage project phase gates.

GRIDALL: A project management framework that supports the development of a comprehensive and connected project management methodology. GRIDALL stands for goals, risks, issues, decisions, actions, and lessons learned.

ICH: International Conference on Harmonisation.

Investigational New Drug Application (IND): The application required to obtain regulatory permission to conduct clinical investigations with an investigational drug in the United States. An IND submission includes technical information on the drug, dosage form, preclinical animal studies, information on investigators, clinical trial protocol, and the Investigator's Brochure.

Indication: A condition or disease that a drug is intended to treat. For approved drugs, it is the condition or disease that is listed on the label.

Investment Opportunity: An endeavor that requires resources (money or effort) that has value to the organization. Investment opportunities can be categorized as "committed" or "uncommitted" based on budget status.

IP: Intellectual Property

Issue: An event that has already occurred and has impacted or is currently impacting a goal. A *risk* converts to an *issue* at the occurrence of a trigger event.

Knowledgebase: An online repository of curated information that can be drawn upon by a user. Helpful for capturing recommended practices.

Life Cycle Team (LCT): The LCT is a cross-functional team with overall responsibility for value-added activities for marketed products.

LRP: Long Range Plan

Market Application Submission (MAS): An application to market a medicinal product that is submitted to a health authority (e.g., FDA in the US, EMA in the EU, MHRA in the UK).

Matrix Management Structure: A business structure in which people are assigned to both a functional group (departments, disciplines, etc.) and projects.

Metrics Quantitative: Measures such as the number of on-time projects. They are used in improvement programs to determine if improvement has taken place or if goals and objectives have been met.

Milestone: An event or delivery date of an output that marks the completion of a critical or major activity group. A milestone is a single point in time; it has no duration or effort.

Mitigation Plan: A set of activities intended to reduce the probability of occurrence or impact of a risk. Compare with *Contingency Plan*.

MOA: Mechanism of action

MTD - Maximum Tolerated Dose

New Drug Application (NDA): The application submitted to the FDA to market a drug product in the United States.

Non-clinical Development: The generation of information, data, and materials primarily related to toxicology, metabolism, bioanalysis, pharmacokinetics, medicinal chemistry, analytical development, etc., that is undertaken to support the development of a compound in clinical development.

Net Present Value (NPV): A calculation used to evaluate the relative merits of two or more investment alternatives. The NPV is calculated as the sum of the total present value of incremental future cash flows plus the present value of estimated proceeds from sale. Whenever the NPV is positive, the investment is generally considered to have merit.

Opportunity: In the context of risk management, it is an uncertain event that can result in a positive impact on a predefined goal.

Parametric Estimating: Estimating using an algorithm in which parameters that represent different attributes of the project are used to calculate project effort, cost, or duration. Parametric estimating is usually used in top-down estimating.

Phase: A grouping of activities in a project required to meet a major milestone by providing a significant deliverable, such as study results. A project is broken down into a set of phases for control purposes. The phase is usually the highest level of breakdown of a project in the WBS.

Phase Gate: Approval points during a project life cycle where a decision is made to move the project to the next phase or to close the project.

Pipeline: The set of programs in the portfolio that are actively being executed. *Compare with Portfolio.*

Pivotal Study: A study intended to support a market application. Pivotal studies are conducted to GCP standards and subject to intensive monitoring to ensure the validity of the data. Compare to *supportive studies.*

POC: Proof of concept

Project: In biopharma, a project is commonly defined as "asset + indication". There is typically one IND per project.

Project Management Methodology: The set of tools, templates, and processes that support the project management activities for an organization.

PMO: Project/Program Management Office.

Project Portfolio Management (PPM): The process by which a set of programs are identified, evaluated, ranked, matched to available resources, and approved for implementation.

Predecessor: A task that must be started or finished before another task or milestone can be performed.

Post-mortem: An approach to capturing lessons learned by reviewing the entire project after it completes. Compare to *continuous improvement.*

Process: A series of steps or actions to accomplish something. A natural series of changes or occurrences.

Project: A group of related work activities organized under the direction of a project manager, which, when carried out, will achieve specified objectives within a stated timeframe. In drug development, *Project* refers to a molecule plus a single indication being sought for that molecule (i.e., NDA or sNDA).

Project Management Methodology (PMM): A system of practices, techniques, procedures, and rules used by those who work in the project management discipline.

Project Manager: The person responsible and accountable for managing a project's planning and performance. The single point of accountability for project operations.

Project Plan: An optimized, detailed, living schedule of all tasks and resources in a project.

Program: A group of related projects addressing a common business goal or problem/opportunity. In drug development, *Program* refers to a molecule and all the indications being sought for that molecule.

Portfolio: All the assets within a company, active or inactive. *Compare with "Pipeline".* Portfolios can also be sliced into pieces (for instance, the Portfolio for Oncology Therapeutic Area).

QA: Quality Assurance

QC: Quality Control

RBS: Resource Breakdown Structure.

Resource: Any tangible support such as a person, tool, supply item, or facility used in the performance of a project. Human resources are people.

Resource Loading: The process of assigning resources (people, facilities, equipment) to a project, usually activity by activity.

Risk: An uncertain event that can result in a negative impact on a predefined goal (also known as a *threat*).

Risk Management: The process of identifying, assessing, responding to, and monitoring risks.

Risk Response Plan: Part of risk management in which planners define actions to be taken for identified risks. Plans can be mitigations or contingencies.

Risk Management Plan (RMP): A document that describes the methodology, process, and tools that will be used to manage the risks for a project.

Sequencing: A part of the scheduling process in which the tasks are positioned serially or in parallel to one another, based on dependencies between them. Sequencing results in a task network.

Subject Matter Expert (SME): Also known as a Technical Expert. An expert in some aspect of the project's content expected to provide input to the project team regarding business, scientific, engineering, or other subjects. Input may be in the form of requirements, planning, resolutions to issues, or reviews of project results.

Sponsor: 1. The company responsible for the initiation, management, and financial support for a clinical trial. 2. Individual(s) with ultimate authority, approval, and accountability for a project.

Stakeholders: Groups or individuals affected by or participating in a project. One with a stake or interest in the outcome of the project. Also one affected by the project.

Stretch Goal: An upside to a goal that can be achieved if additional resources are applied (e.g., adding a study startup unit to achieve FPI sooner).

Successor: A task or milestone that is logically linked to one or more predecessor tasks.

Supportive Study: A clinical trial that is intended to support the conclusions of a pivotal study in a market application submission. Compare with *pivotal study*.

Task: An activity that has been assigned to a person. *Compare with Activity*. A task is the lowest level in the WBS.

Target Product Profile (TPP): The statement of requirements and desired characteristics for an asset that enters the market. TPPs are set at the indication level and are used to design the clinical development plan for that indication.

Variance: The difference between estimated cost, duration, or effort and the actual result of performance. In addition, it can be the difference between the initial or baseline project scope and the actual deliverable.

Work Breakdown Structure (WBS): A tool that defines the hierarchical breakdown of all work in a project. It is created by the process of decomposition of high-level goals into discrete, actionable activities.

Work Package: The work defined at the lowest level of the work breakdown structure for which cost and duration can be estimated and managed. This is the level at which accounting is performed for budget vs. actual comparisons.

Workaround: A response to a threat that has occurred (i.e., issue), for which a prior response had not been planned or was not effective.

Editor

Dr. Joseph P. Stalder, Pharm. D., PMP, is the founder of Groundswell Pharma Consulting, based in San Diego, USA. He received his Bachelor's degree in Pharmacology from the University of California, Santa Barbara, and his Doctor of Pharmacy program at the University of California, San Diego. After receiving his Pharm. D., he completed a post-doctoral fellowship at Forest Laboratories in New York/New Jersey, during which he also served as an adjunct professor at St. John's University in Queens, NY. Joe has more than 10 years of project management experience in pharmaceutical development in both large and small companies. His experiences include PM Department Head, PMO Head, and PM Lead on several late-stage development assets in oncology, infectious disease, cardiology, metabolism, and pulmonology.

Contributors

Mark Christopulos
Strategy Execution Consultant

Henri Criseo
Director, Project Management
Global R&D
Santen

Kamil Mroz
PMI, PMI-ACP, DASSM, Agile Coach
Program Management Lead
UCB

Norbert Leinfellner
Head of PMO
Element Science, Inc.

Hourik Miller
Clinical Development Business Strategy and Operations
Gilead Sciences

Rhonda Peck
Senior Drug Development Scientist
Cook Biotech, Inc.

Dave Penndorf
Life Science PPM Practice Head
Planisware

Courtland (Corky) LaVallee
Global Program Team Director
EQRx, Inc.

Part 1

The Role of PM in Drug Development

1 Overview of Drug Development

Hourik Miller
Gilead Sciences

CONTENTS

1.1 INTRODUCTION

The development of biopharmaceutical products is a complex, risky, and expensive endeavor. A typical drug development project can take 12–15 years from candidate identification to market launch. Along the way, hundreds of people are involved, from dozens of technical backgrounds. Several hundreds or thousands of tasks and deliverables need to be coordinated, many times with both short-term and long-term considerations that need to be strategically evaluated. On top of that, the drug development process itself is dynamic and continuously evolving. Despite all the complexity and variation possible, there are highly defined and well-established stages and milestones that apply to all drug development projects, and a project management professional in the biopharma industry needs to be fully aware of the basic requirements for each of these stages. In this chapter, we will cover the basic principles and deliverables of each phase of development in the United States that a project management professional will encounter.

DOI: 10.1201/9781003226857-2

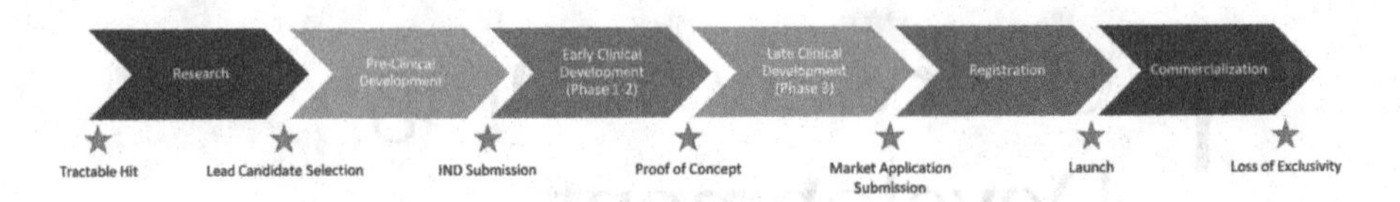

FIGURE 1.1 Overview of the phases and stage-gates of a traditional drug development lifecycle.

A high-level schematic of the drug development process is shown in Figure 1.1. During the research stage, many novel compounds are synthesized and tested for desirable chemical attributes. After a lead candidate is identified and selected for preclinical studies, a battery of tests on that compound are conducted in a variety of in vitro and animal models to predict how the compound will perform in humans. The data from those studies are summarized into an investigational new drug (IND) application, the first major milestone of the process.

Successful clearance of the IND through the FDA allows the sponsor to begin studying the compound in humans, a stage termed clinical development. During this state, drug candidates go through a carefully regulated and managed process to determine their actual effect on humans across a wide range of parameters. Throughout all phases of drug development, the ability to formulate these compounds such that they are shelf-stable, can be manufactured in a consistently reproducible fashion, and can be administered accurately and as conveniently as possible is constantly being assessed and improved.

Traditional clinical development paradigms divide this stage into early clinical development, which includes Phase 1 and Phase 2 trials, and late clinical development, which includes the pivotal Phase 3 trials. If data from the pivotal studies support a market application, the data from the entire development plan are summarized into a new drug application (NDA) and submitted to the FDA. The FDA then reviews the application during the registration stage.

If the FDA approves the drug for commercial use, the sponsor then launches the drug and begins commercialization. Commercialization continues until the sponsor decides to cease investment and marketing activities, typically at the time the asset loses market exclusivity.

1.2 PRECLINICAL DEVELOPMENT

The purpose of preclinical development is to move a lead compound from laboratory and animal studies into human studies. As shown in the schematic, the output of this phase is an IND application submission, which includes a comprehensive body of evidence intended to justify the safe first administration to humans. One often hears the terms "preclinical" and "nonclinical" used synonymously to describe studies not performed in humans; however, in this book, we define preclinical to mean the subset of nonclinical studies that are required before human studies can be conducted and that specifically support the decision to advance an IND candidate into human trials. This subset of studies is also referred to as "IND-enabling studies". There are usually additional nonclinical studies that are conducted *after* a compound enters the clinic that further characterize the compound (e.g., long-term toxicity) or support broader development (e.g., juvenile animal studies before the drug can be administered to pediatric patients).

Generally speaking, IND-enabling studies include in vitro (outside of a living organism) and in vivo (inside of a living organism) assessments that characterize the pharmaceutical, pharmacological, and toxicological properties of a drug. These studies include the following:

- Pharmacodynamics (PD) (pharmacological studies which define the biochemical and physiological effects of drugs, i.e., what the drug does to the body)
- Pharmacokinetics (PK) (pharmacological studies that define drug properties in the context of absorption, distribution, metabolism, and excretion, abbreviated as ADME, i.e., what the body does to the drug)
- Animal toxicology (2-week to 12-week-long studies that characterize drug toxicity in a variety of animal species)
- Mechanism of action (efforts that identify the specific biochemical interactions caused by a drug in order to produce a desired effect)
- Bioanalytical methods (efforts that establish the procedures involved in the collection, processing, storage, and analysis of the drug)

Advancing a drug into initial Phase I trials requires completing numerous nonclinical studies including, but limited to, those mentioned above. These studies must comply with all detailed regulatory guidelines, such as Good Laboratory Practices (GLP) or guidance as per the International Council for Harmonization (ICH), and convince regulatory agencies, oversight committees, and review boards that a drug is suitable for administration in humans.

Later, Phase 2 and 3 clinical development and regulatory approval for marketing require additional nonclinical studies, for example, longer-duration toxicology and reproductive toxicology, carcinogenicity, and biodistribution studies. Special studies may also be needed to address concerns unique to a specific drug or its molecular target and mechanism of action.

For further reading, consider the relevant regulatory guidance documents

- ICH M3(R2) "Nonclinical Safety Studies for the Conduct of Human Clinical Trials and Marketing Authorization for Pharmaceuticals"
- ICH S6 and S6 Addendum "Preclinical Safety Evaluation of Biotechnology-Derived Pharmaceuticals".

In the following sections, we will further describe the key concepts of preclinical development mentioned above.

1.2.1 Pharmacodynamics (PD)

Pharmacology plays a key role in animal models. IND-enabling safety pharmacology studies assess the effects of a drug on the cardiovascular, central nervous, and respiratory systems in animals. Primary PD studies are generally also included to assist in obtaining PK–PD relationships, to test compounds for efficacy, develop and define biomarkers, and ultimately define the therapeutic effects of the drug, including relationships to dose and/or exposure. Regardless of how the drug effect occurs, ultimately it is the concentration of the drug at the desired site of action that determines whether or not the drug has an effect. This quantification of a drug's effect at the molecule, interaction, organ, and organism levels is the output of PD studies. PD endpoints are outcome measures that capture the potential clinical benefit of a medicinal product and vary depending on the hypothesis of the study. Examples of PD endpoints include measurements through age-related scoring assessments, blood draws, or ECGs.

1.2.2 Pharmacokinetics (PK)

PK can make or break a molecule or compound. IND-enabling PK assessments typically include in vitro metabolism and plasma protein binding studies, as well as systemic exposure studies in the same species as repeated-dose toxicity evaluations. While these studies are generally sufficient for supporting initial human trials, additional studies to characterize the ADME profile of a drug are often required prior to conducting later phase clinical studies. The amount of information needed depends in part on the phase of development and design of the proposed clinical study. PK endpoints are outcome measures that capture how the drug is absorbed, distributed, metabolized, and excreted in the body. Examples of PK endpoints include maximum concentration (Cmax) and area-under-the-concentration curve (AUC).

1.2.3 Toxicology

Toxicology plays a critical role across the entire pharmaceutical product lifecycle starting from preclinical studies through post-marketing studies. There are guidance documents available to help researchers and drug developers know what studies need to be conducted to support clinical trials, such as the ICH M3 (R2), which details the considerations for the design of nonclinical toxicology programs supporting the clinical development of small molecule drugs for non-oncology indications.

There are many types of preclinical toxicology studies, and the decisions on what studies are needed in which cell lines or animal species are a key part of strategic planning in this area. Note that while some preclinical studies do not have to be performed using GLP, any preclinical study assessing safety and intended for regulatory authority submission in support of a human clinical trial must adhere to those practices, as well as applicable standards and guidance issued by the appropriate regulatory authority (e.g., Food and Drug Administration in the United States), the International Council for Harmonization (ICH), and the International Organization for Standardization (ISO). Animal toxicology studies should use the same route of administration as is planned for humans, and the drug tested should be as similar as possible to the clinical material. However, the drug used in animal studies does not need to be manufactured according to Good Manufacturing Practice (GMP) standards.

A program of toxicology and pharmacology will be completed to allow the assessment of doses of the drug that can reasonably be tested on humans. The level of drug exposure achieved in the toxicology studies will be determined and related to planned human exposure levels. Any findings from the toxicology and pharmacology studies will be carefully assessed to decide whether the drug can, with reasonable safety, be administered to humans and to ensure close monitoring of potential clinical symptoms related to preclinical findings.

There are three primary systemic toxicology tests that occur in the preclinical setting, as described below:

- **Acute toxicity:** Acute toxicity studies are generally conducted in two mammalian species (one rodent and one non-rodent) using the clinical route of administration and a parenteral route (e.g., intravenous or subcutaneous). They define the dose range associated with toxicity and should be chosen for studies that will allow the determination of a maximum tolerated dose (MTD) and a no-observed-adverse-effect level (NOAEL) in humans. These parameters are important for predicting human safety and for clinical dose selection. Acute toxicity studies can be the primary IND-enabling toxicology studies, but they are often combined with repeated-dose toxicology studies.
- **Subacute toxicity:** Subacute toxicity studies of 14 or 28 days of repeated drug exposure are intended to reveal initial toxic effects. These studies are used in comparison with single-dose studies to indicate the potential for accumulation of the drug. As a general rule, repeated-dose studies are designed with duration and route of administration similar to those of the proposed clinical trial. Dose levels and dose regimens should be selected so that observed exposures in nonclinical species will adequately cover expected clinical exposures.
- **Subchronic/chronic toxicity:** Subchronic or chronic toxicity studies of up to 6 months in rodents and 9 months in non-rodents are intended to reveal targets for longer-term toxic effects and to define doses associated with adverse effects and NOAEL.

In addition to systemic toxicity studies, there are several descriptive toxicology studies involved in preclinical development. Some of the more common descriptive toxicology studies are described below:

- **Genotoxicity:** in vitro and in vivo tests designed to identify whether the drug induces damage to genetic material either directly or indirectly by various mechanisms.
- **Mutagenicity:** these studies include a variety of in vitro tests that define any potential effect on DNA that may be linked to different types of toxicity, such as carcinogenicity. To determine the mutagenic potential of an investigational drug, a gene mutation assay (e.g., Ames assay) is conducted to support single-dose clinical trials. Additional chromosomal damage assessments are conducted to support repeated-dose clinical studies. Complete genotoxicity testing should be completed prior to Phase 2.

- **Carcinogenicity:** these studies are typically 2-year rodent studies using repeated doses given throughout the lifetime of the animal. Long-term carcinogenicity studies are typically carried out after the initial IND submission.
- **Immunogenicity:** specific to biologics, immunogenicity studies determine whether the species' own antibodies are raised against the drug, which can lead to neutralization, binding, sustaining, or cross-reacting antibodies with endogenous compounds.
- **Immunotoxicity:** these studies determine whether the drug induces inadvertent side effects with regard to autoimmunity (hypersensitivity or allergy) or immunosuppression.
- **Developmental/reproductive toxicity:** These studies use repeated doses given prior to mating and throughout gestation in order to assess the effect of the drug on fertility, implantation, fetal growth, fetal abnormalities production, and neonatal growth.

In addition to characterizing the pharmacological and toxicological properties of the lead compound, the preclinical stage also involves Chemistry, Manufacturing, and Controls (CMC) activities focused on developing an initial formulation and generating enough drug to support preclinical evaluations. These include optimizing the drug substance (the active pharmaceutical ingredient, or API) manufacturing process, development of analytical methods to enable assessment of the purity of the API, stability of the drug product under short-term storage conditions, and scaling up production to support clinical studies.

All of the preclinical and CMC data are included in an application that is submitted to the appropriate regulatory authority (e.g., an IND to the FDA in the United States, CTA in other countries). The preparation of such a submission can follow many of the same principles of a market application submission described in Chapter 13.

1.3 EARLY CLINICAL DEVELOPMENT

Upon receiving a notification from the regulatory body to proceed to clinical trials, (typically 30 days after the IND submission to the FDA) and barring any clinical holds, drug developers can initiate early clinical development.

There are strategic decisions that need to be made before setting the early clinical development plan, and it is important for the project management professional to understand the advantages and disadvantages of each option to help guide the team to an appropriate strategy. For example, clinical pharmacology studies can be staged in a way to inform the design of later-stage studies that saves money and human exposure but at the same time making the later-stage studies as efficient and informative as possible.

- A study of the effect of food study on PK can be conducted in early development so that recommendations can be made about the timing of drug dosing in relation to mealtimes for the later-stage studies.

- Drug interaction studies may also be conducted in early development to determine whether concomitant medications that are common for the planned population can be allowed in the later-stage studies.

1.3.1 Phase 1

Phase 1 studies aim to demonstrate the safety, tolerability, PK (what the body does to the drug), and PD (what the drug does to the body) of an investigational drug in humans, with the goal of providing data to advance the candidate into later-stage studies. Phase 1 studies often involve a small number (usually <50) of healthy volunteers and typically do not have an efficacy-demonstrating component to their designs. Some Phase 1 protocols may include secondary or exploratory endpoints such as biomarker assessments. In some cases, such as with oncology or other life-threatening diseases, the enrollment of healthy volunteers would be considered unethical due to the toxic nature of the drugs being evaluated; therefore, patients with the disease are enrolled and monitored not only for safety but also for efficacy endpoints.

Phase 1 trials include the first-in-human trial and food effect trials that are typically done in early development as well as other clinical pharmacology trials that are done later in development. Figure 1.2 outlines some of the Phase 1 trials that are common for a clinical development plan, as well as some of the preclinical studies that are used to inform the design of the clinical trials.

Typically, the first-in-human Phase 1 study will be a combined single ascending dose (SAD)/multiple ascending dose (MAD) study. In the SAD portion, a group of subjects receives one dose of the study drug and is then monitored for potential adverse effects and analyzed for PK parameters. The next group of subjects receives a larger dose and so on until subjects start to exhibit adverse events (AEs) of sufficient severity to require limiting exposure to larger doses (dose-limiting toxicities or DLTs). The starting dose is determined by the preclinical toxicology studies described in the previous section. In the MAD portion, subjects receive multiple

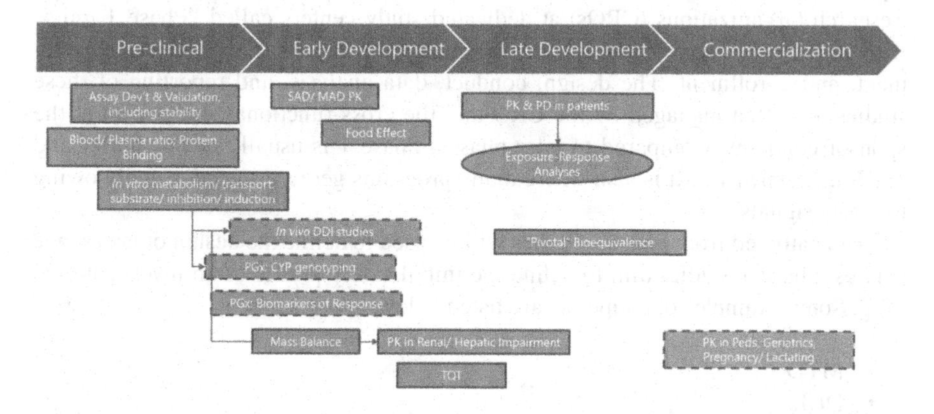

FIGURE 1.2 Common clinical pharmacology activities throughout the drug development process.

doses of the study drug. In most cases, the output of this Phase 1 trial will be a maximum tolerated dose (MTD), defined as the highest dose of a drug that can be administered without causing unacceptable side effects that cause more risk than benefit to humans. However, recently the FDA, through an effort termed Project Optimus, has challenged this MTD paradigm and pushed for more appropriate determinations of a recommended Phase 2 dose.

Other Phase 1 studies used to evaluate the clinical pharmacology of the drug include the following:

- **Food effect:** These studies are conducted to assess whether food affects the rate and extent of drug absorption. Drug is administered with food, without food, or with certain types of food (such as high-fat food). The results of these studies are likely to contribute to food-related language on the drug label.
- **Human mass balance (aka hAME)**: These studies are conducted to understand how a drug is absorbed, metabolized, and excreted.
- **Renal or hepatic impairment**: These studies are conducted to characterize the effect of kidney and liver function on a drug's PK profile.
- **TQT**: These studies are conducted to determine whether a drug could cause heart arrhythmias. In a typical thorough QT (TQT) study, healthy volunteers are administered the drug and their heart activity is monitored by an electrocardiogram.
- **Special populations**: These studies are conducted to determine whether the drug's PK differs in various populations, such as women (pregnant and nonpregnant), children, and the elderly.
- **Bioavailability**: These studies are conducted to determine the fraction of the drug that is absorbed in the blood. The higher the bioavailability of the drug is, the more efficacious it could be. In addition, a drug with higher bioavailability can potentially cost less to manufacture.

Operationally, Phase 1 studies can be conducted by sponsor-chosen Contract Research Organizations (CROs) at dedicated study centers called "Phase 1 units" with staff that are experienced in first-in-human or Phase 1 trial conduct, management, and enrollment. The design, conduct, data analysis, and reporting of these studies are often managed by the CRO and the cross-functional study team at the sponsor company. Compared to later phases, Phase 1 is usually much shorter and much less expensive. It is also where many programs get terminated, usually owing to safety signals.

Data gathered from Phase 1 studies will be used to define the design of later-stage studies. Phase 1 studies aim to define the initial safety profile of an investigational drug. Some examples of endpoints are listed below:

- MTD
- DLTs
- Incidence of adverse events or side effects
- For oncology/life-threatening diseases: progression-free survival (PFS) or overall survival (OS)

1.3.2 Phase 2

After determining the Phase 2 dose from the first-in-human SAD/MAD trial, Phase 2 clinical trials are conducted to further define the safety profile of the drug in patients with the target disease and to further evaluate PK parameters. In particular, the dose–response relationship is evaluated to support the selection of appropriate doses and dose regimens that will be tested in Phase 3 pivotal trials. In addition to continuing to look at safety and tolerability, Phase 2 trials also often include endpoints to assess the preliminary efficacy of the drug on defined disease outcomes; hence, they are often referred to as proof-of-concept (PoC) studies.

Using data from the Phase 1 first-in-human SAD/MAD trials and from preclinical in vitro and in vivo studies, three or four doses or dose regimens are selected for the Phase 2 trial. In some therapeutic areas, useful "markers of activity" may be measured in Phase 1 or Phase 2a trials, which would assist in the selection of doses.

The size, duration, and cost of Phase 2 trials may vary considerably depending on the therapeutic area and disease state. For example, a Phase 2 trial for a drug with a dependably consistent endpoint and a drug with a strong efficacy margin may require only 25 patients per dose group. In contrast, a trial for a drug with a highly variable endpoint and a drug that is predicted to have a small incremental treatment effect may require a 1500-patient trial.

Drug sponsors and regulators will use data from the Phase 2 study to identify the best dose and dose regiment for the pivotal Phase 3 trial. The chosen dose may be the mid-dose of a Phase 2 trial because no significant improvement in efficacy was offered by the high dose. Alternatively, the mid-dose may be selected because the high dose provided marginal improvement in efficacy but significantly increased the incidence of adverse effects, resulting in an overall worse benefit-to-risk assessment at the high dose.

On the CMC front, during early development, the drug's manufacturing process will ideally have been optimized and "locked". This allows for a "commercially representative" API to be used in the long-term toxicity studies and for the Phase 3 trials. In particular, it is important that the impurity profile is consistent between the material used in late development and what will be used for commercialization.

1.4 LATE CLINICAL DEVELOPMENT

After the optimal dose and dose regimen are identified from the Phase 2 studies and assuming sufficient efficacy is observed in the Phase 2 trial to justify further development investment, the drug developer will move to late clinical development where pivotal Phase 3 trials are run. In contrast to early clinical development, there is less room for creative strategies in this stage. The pivotal trials are intended to confirm that the drug at the selected dose and dose regimen is adequately safe and effective in treating the disease. The result of the trial is a binary outcome: it either achieved the objective or it did not.

1.4.1 Phase 3

The objective of Phase 3 trials is to confirm the efficacy of the investigational drug and continue to collect information that allows the drug to be used safely, including data on potential long-term effects of the drug. Phase 3 studies are conducted on larger groups of people (at least several hundreds and often in the thousands) and can take many months to a few years to complete. Some Phase 3 studies may include placebos, but the placebo is never used alone if there is a marketed treatment that works. In this case, the patients who receive a placebo are offered the standard of care treatment as well. Because Phase 3 trials are conducted in the diseased population, the product-label intent needs to be clearly reflected in the study protocol to ensure that the drug does what it is intended to do and that the product label or insert appropriately reflects that. The commercial intent, i.e., if the drug offers superior efficacy to a marketed competitor product, or if the drug demonstrates noninferiority to the competitor but has other benefits, must also be aligned with the hypothesis. The trial hypothesis will have an important influence on the design of the study and the scale of patient recruitment and hence duration and costs. For chronic therapy drugs, it will be expected that long-term drug exposure will form a key safety component of the dossier with significant numbers of patients (e.g., ~500) dosed for 6–12 months. The analysis and report of the vast amount of clinical data needed to create the clinical registration documents may take six months from the end of patient dosing.

Operationally, Phase 3 studies are also more likely to be offered in local community hospitals and doctor's offices, in many places across the country and even the world.

The output of these trials is a dossier that includes all the data required to demonstrate a clear benefit-to-risk justification for the use of a drug in a defined patient group for a specific clinical intent. All of the data pertaining to the drug gathered from preclinical/animal studies and human trials are integrated into an NDA and submitted to the FDA for approval. In Europe, the application is called the Marketing Authorization Approval and is submitted to the EMA, which is the main regulatory agency equivalent to the FDA.

1.5 INNOVATIVE CLINICAL DEVELOPMENT STRATEGIES

The above description of a clinical development path might be described as "traditional", but there are also innovative approaches that may support registration that are faster and more cost-efficient. These innovative approaches are commonly employed in therapeutic areas such as oncology and rare diseases. Many innovative strategies have been successful with the FDA, and we will highlight a few below.

1.5.1 Seamless Trials

Clinical trials that combine phases often result in faster data readouts because of the operational efficiencies gained by keeping study sites open to enroll patients. Common use cases are Phase 1/2 or Phase 2b/3 trials or, as with the COVID trials, Phase 1/2/3 trials. Referred to as adaptive, or seamless, clinical trials, these types of trials allow for modification to the protocol based on emerging data at predefined interim analyses during the trial. Learnings from the available data can be applied to the design of the rest of the trial, such as adding or removing treatment arms at a certain dose level or cohorts of a certain disease state or adaptation to the randomization schema, stratification factors, or eligibility criteria. The goal is to proceed with enrolling additional patients into the treatment arms demonstrating the highest potential for successful outcomes. Note that any adaptation must be clearly defined before unblinding any data (in the case of blinded clinical trials) so as to not compromise data integrity and validity, and in most cases, a data safety monitoring board is relied upon to protect the data and prevent bias.

1.5.2 Master Protocols

Recognizing the challenge of conducting trials with targeted therapies in subsets of populations, there is increasing interest in conducting trials using master protocols. The master protocol design allows multiple interventions to be tested in multiple diseases. They can be categorized as umbrella trials, basket trials, and platform trials.

Type of Trial	Objective
Umbrella	To study multiple targeted therapies in the context of a single disease
Basket	To study a single targeted therapy in the context of multiple diseases or disease subtypes
Platform	To study multiple targeted therapies in the context of a single disease in a perpetual manner, in which therapies are allowed to enter or leave the platform on the basis of a decision algorithm

Master protocols require more planning than traditional trials, but the potential to quickly translate laboratory findings to clinical evaluation is very encouraging as drugs become more and more precise. An article published in the New England Journal of Medicine in 2017 describes master protocols in detail, and if your team is considering this type of study, I recommend diving deeper into that article[1].

1.5.3 Regulatory Flexibility

Many sponsors of drugs to treat life-threatening diseases with high unmet medical needs may benefit from regulatory flexibility around the robustness of evidence to support market approval. Typically, regulatory approval requires data from two well-designed randomized controlled trials that often require several hundred patients and spans many years. However, this development approach withholds therapy from patients until after Phase 3 trials are completed, raising ethical concerns about equitable access to life-saving medicine. To expedite access, some regulatory agencies have adopted expedited processes that allow for a less mature data package (e.g., data from single-arm Phase 2 or even Phase 1 trials) to be submitted for review. As written in the FDA's draft guidance *Demonstrating Substantial Evidence of Effectiveness for Human Drug and Biological Products*, there are circumstances where the Agency allows flexibility to the traditional clinical development approach.

[1] Master Protocols to Study Multiple Therapies, Multiple Diseases, or Both | NEJM.

The guidance explicitly describes development for diseases that are life-threatening or severely debilitating with an unmet medical need and diseases that are rare.

Many health authorities have adopted expedited approval paradigms that conditionally approve a drug for public use while the rest of the development activities are completed. In the United States, the FDA has an accelerated approval pathway for drugs that treat serious conditions that can use surrogate endpoints as a predictor of clinical benefit. A conditional approval will allow the sponsor to market the drug in parallel to conducting confirmatory clinical trials that will provide the typical robust data package.

1.6 REGISTRATION

The registration phase for a project encompasses the period from market application submission to regulatory approval for the drug to be marketed. The review period varies depending on the health authority. In the United States, the review periods are 10 months for standard review and 6 months for priority review. In addition, the FDA currently has the following four expedited programs that are intended to accelerate drug development, data review, and approval:

1. Fast Track Designation – expedites development and FDA review of drugs to treat serious conditions where there is a significant unmet medical need
2. Breakthrough Therapy Designation – expedites development and FDA review of drugs that may demonstrate substantial improvement over available therapy
3. Priority Review – indicates that the FDA will take action on an NDA within 6 months (compared to 10 months under standard review)
4. Accelerated Approval – allows drugs for serious conditions, where there is an unmet medical need, to be approved based on a surrogate endpoint

During the registration phase, the clinical program generally continues to run with additional trials being conducted, which are intended to provide data to support the marketing of the product (phase 3b marketing studies) or studies exploring new indications for the drug. Non-registrational data generation (NRDG) activities also occur during this period.

1.7 LIFECYCLE MANAGEMENT

During the Lifecycle Management (LCM) stage, additional investments are made to maximize the commercial value of the product. These investments may include

- Development in related indications (e.g., a similar type of cancer) to expand the label to a broader patient population or to address entirely different diseases
- Expansion of current indications into new populations, such as pediatric, geriatric, or hepatically or renally impaired patients

- Conducting comparative studies or NRDG, including real-world evidence, and research to further differentiate the product from other marketed drugs
- New formulations or dose regimens – either bioequivalent (e.g., converting to a more shelf-stable salt) or non-bioequivalent (for example, an extended release that improves the tolerability profile of the drug). Non-bioequivalent formulations are treated as new chemical entities by regulatory agencies.

For a major brand, LCM can end up costing much more than the initial development investment but with substantial benefits. A sponsor will attempt to demonstrate the benefits of the product to "raise the bar" on future competitors. For example, the first mover in an oncology indication will always have the advantage of a more mature dataset to demonstrate OS at longer time points, so the sponsor will continue to invest in generating, publishing, and promoting this unique point of differentiation as a way to defend the franchise.

In addition to increasing revenues, LCM trials also provide the opportunity to learn more about the drug in a large patient population. For example, the patient-safety database can start to include tens of thousands or hundreds of thousands of patients, allowing the sponsor to identify rare adverse drug effects that may not have been detectable during the clinical development program where only a few hundred or thousand patients were studied.

LCM investments continue throughout the life of the product until a decision is made to discontinue its promotion. This decision often relates to the loss of exclusivity from patent expiry or loss of regulatory protection. At this point, an innovator company will usually divest the product to a company that will continue commercialization during generic competition.

1.8 SUMMARY

Ultimately, a safe and effective drug is the outcome of successful drug development, a process that can take many years if not decades from start to finish. Drug development requires major investments over a long timeframe in a highly regulated environment to bring a drug to market and maximize its market potential. There are multiple phase gates and decision-enabling milestones throughout the process to ensure all necessary considerations and steps are taken prior to moving a drug candidate forward. Innovative, adaptive, and agile approaches can highlight risks and potential failures earlier in the process, thereby reducing cost, and preventing unnecessary adverse effects from occurring in both healthy and disease populations.

Development projects are exceedingly complex ventures that can involve unprecedented scientific approaches to often intractable diseases and thus are very risky. It is a truism in the industry that far more projects will fail than will succeed, especially in the earlier phases of human clinical trials. Project management in drug development requires a foundational understanding of the key principles that drive decision making for a potential new drug, or new use of an existing drug. Excellent project management can substantially reduce any risks associated with new drugs and increase the likelihood of a drug candidate's success.

2 Project Management's Place in Drug Development and the Various Project Management Roles in Biopharma

Joseph P. Stalder
Groundswell Pharma Consulting

CONTENTS

2.1 INTRODUCTION

As biopharma companies shifted in the late 1990s from independently managed functional activities to centrally managed cross-functional projects, the need to coordinate and integrate these cross-functional teams and work streams has given rise

DOI: 10.1201/9781003226857-3

to having dedicated project management practitioners. Despite being a late adopter of project management, the biopharma industry has quickly embraced the systematic and formalized approach that project managers bring to planning and executing complex projects. Nowadays, project management has become a standard practice in the biopharma industry, and most biopharma leaders today would consider Project Management as a core business function just as much as Chemistry, Clinical Science, or Regulatory Affairs.

Not only has the application of the project management discipline become standard to the clinical development stage, but it is also now being applied to many other stages of drug development (e.g., preclinical, commercial) and even to specialized functions within clinical development (e.g., Regulatory, Chemistry, Manufacturing and Controls (CMC), Medical Affairs). Furthermore, while the standard aspects of planning, coordination, scheduling, monitoring, and managing the risks of various drug development activities is still appreciated, the role of project managers has also expanded into areas of project leadership, team effectiveness, governance, strategic value optimization, process improvement, and more.

This chapter describes the evolution of the project management function in the biopharma industry, some of the current roles and responsibilities that project managers fulfill, and the skills and competencies that project managers need to be effective in biopharma.

2.2 WHAT IS PROJECT MANAGEMENT AND WHAT DOES A PROJECT MANAGER DO?

There are many ways to describe what project management is, and none are complete or always accurate for every company. In general, project management is concerned with how a project will run. This can include such things as

- Scope management
- Timeline management
- Cost management
- Risk management
- Resource management
- Team management
- Stakeholder management
- Process management
- Value management

Project management can be viewed as one of the three components, along with appropriate project governance and line management disciplines, that are necessary to successfully deliver a project. In this construct, project management is responsible for delivering projects using resources (staff and funds) provided by functional line management, which is committed to the project by a governing body. This construct has arisen as biopharma companies have moved toward a matrix organizational structure where project team members are drawn from functional departments, in effect creating lines of accountability both to the department or function and to the project. Overseeing and aligning the needs of projects with those of departments or functions is the role of the governing bodies (Figure 2.1).

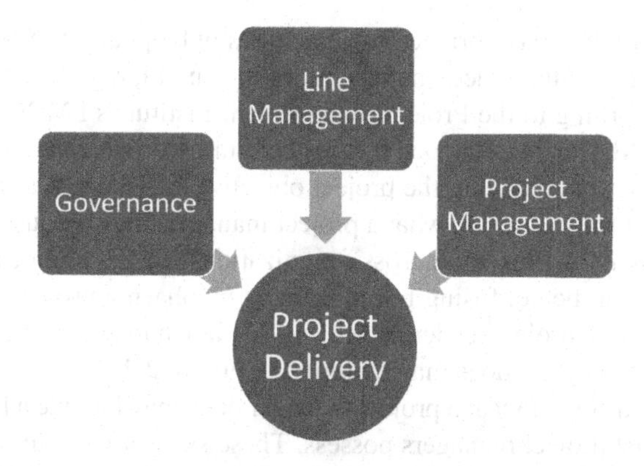

FIGURE 2.1 Three components of effective project delivery: Governance, Line Management, and Project Management

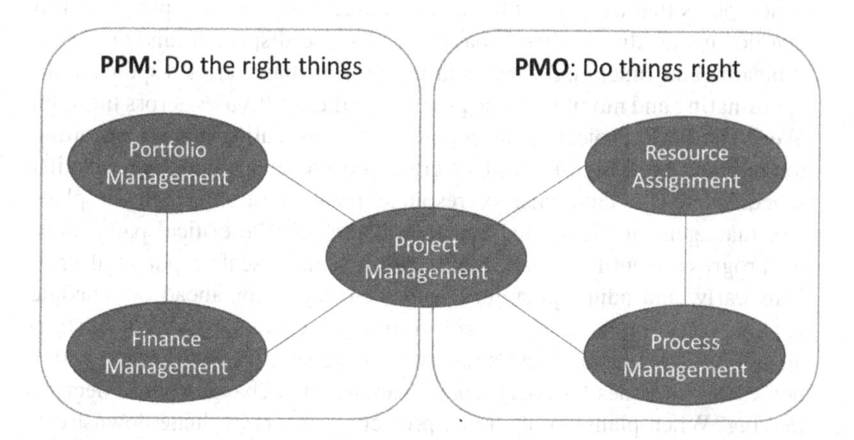

FIGURE 2.2 Conceptual representation of Project Management as the interface of project planning via Project Portfolio Management and project execution via the Project Management Office.

Another way to define project management is in the context of project planning (project portfolio management, PPM) and project execution (project management office, PMO). The project management function straddles both disciplines and interfaces with other functions that support both PPM and PMO activities (Figure 2.2).

These definitions still do not cover all the things a project management function will do across all companies. For example, at a small company, the project management function may also serve as the bridge between two partners in a strategic collaboration. Hence, rather than trying to define what project management is for a biopharma company, it is more accurate to say that project management can provide value whenever the coordination of a complex set of activities is needed to deliver an output or outcome.

Before we get into what a project manager does in biopharma, it is helpful to step back and consider what a project manager does in general, regardless of the industry he/she is in. According to the Project Management Institute's PMBOK (7/e), a **project manager** is "the person assigned by the performing organization to lead the team that is responsible for achieving the project objectives". However, a single definition like this does not fully describe what a project manager does because the role varies so widely across industries and across organizations. In particular, for biopharma, the term "lead" can be confusing because many biopharma projects have a project manager as well as a project leader, as we will see in Chapter 3. A better description of what a project manager does may be found in Figure 2.3.

A further definition of what a project manager does may include a list of skills and competencies that project managers possess. These skills are not unique to a project manager, but instead, it is the unique combination and focused application of these skills that makes a project manager valuable to a project team and to an organization.

- **Plan the Work**: Project managers convert broad corporate strategies into action plans that the project team can deliver. "Big-picture" plans are broken down into discrete work packages that are dispersed among various subteams and functional lines, and the project manager is responsible for coordinating and monitoring the progress of these activities across the team.
- **Work the Plan**: Project managers oversee the execution of work according to the plan. After high-level plans are approved, they create more detailed schedules (e.g., Gantt charts), resource plans, cost management plans, risk management plans, etc., and they determine the critical path, monitor progress, identify variances in cost and time, escalate potential problems early, and adjust plans to ensure delivery on or ahead of schedule. As the project progresses, the project manager evaluates the project status against the broader business strategy to make sure they still align, and, if not, escalates issues for governance awareness and change control decision making. When plans go off track, project managers evaluate downstream

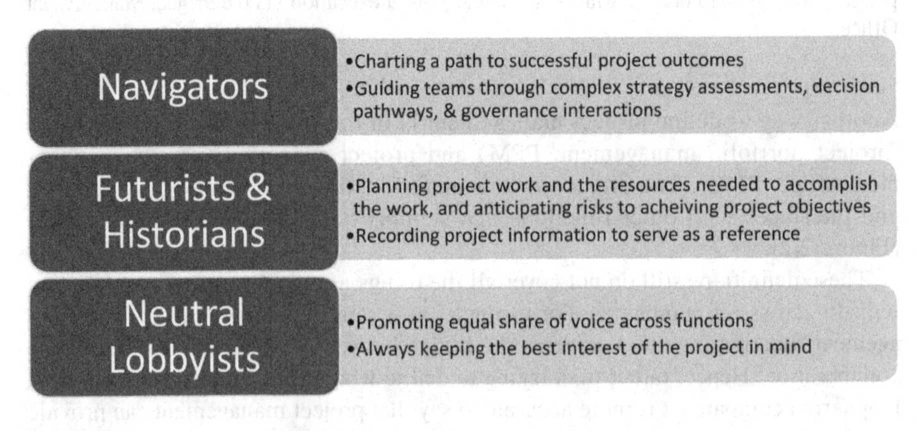

FIGURE 2.3 Laypersons definition of what a project manager does.

implications and build and analyze scenarios, including cost requirements and resource needs, to get the plan back on track.

- **Manage project risks**: Project managers keep a watchful eye on things that could disrupt the project plan. The project manager solicits concerns from team members and then works with the team to assess risks, evaluate response strategies, and implement plans to avoid, accept, mitigate, or transfer the risk. The project manager serves as the conduit for communicating risk response strategies to stakeholders so that they can be aware of and potentially address the risk.
- **Manage the flow of project information**: The project manager is the unifying point of project information and serves as the central point of contact for project team members and management when it comes to operational information. In addition, with their comprehensive, long-term view of the project's goals, they help the team understand how the discrete work packages and short-term activities fit into the big picture. As such, the project manager needs to manage information exchanges across the project team system and up to senior management. This includes preparing presentations, updating systems, and generating progress reports for the project team, functional management, and governing bodies.
- **Manage team dynamics**: Project managers are the glue that keeps project teams together. Whereas project leads are outward-focused, keeping external stakeholders informed of project progress, project managers are inward-focused, keeping the project team's health and performance front of mind. To this end, the project manager helps define the project team structure and ways of working, including evolving the team as the project moves through different stages of development. The project manager needs to define how the team members will interact with each other through charters and meeting management, as well as maintaining team health through team-building activities and conflict resolution between functional lines when needed. The project manager also needs to be aware of issues with team members that could put the project at risk and work with the functional manager to address the issues. Chapter 3 describes in detail the responsibilities a project manager takes on in managing team dynamics.
- **Implement best practices:** As described in Chapter 4, the project manager has responsibilities to the department that go beyond direct project support. They also contribute to the PMO's remit of having a single set of project management standards, tools, processes, and methodologies, and they work to ensure that these best practices are being employed across all project teams.

As noted briefly above, it is important to note that in biopharma, unlike in some other industries, project managers are typically distinct from project leaders. In some industries, the terms "project manager" and "project leader" are synonymous, referring to an individual who is given decision-making authority over a budget and a set of resources to deliver an output or outcome. However, in biopharma, team leadership is usually assigned to two individuals: a project leader and a project manager. We will describe these roles more in Chapter 3.

2.3 EVOLUTION OF PROJECT MANAGEMENT IN BIOPHARMA

The biopharma industry was a relatively late adopter of the project management discipline. Considering that many "modern" project management practices were introduced in the 1950s in the fields of engineering and construction and later refined and further developed in the software and manufacturing industries in the 1980s, the introduction of project management to the biopharma industry really started to take place in the late 1990s.

During the first decade or so after its introduction, there were widely varying ideas of where Project Management fits in a biopharma company. First, the reporting line varied, with many advocates thinking the function should report directly to the CEO; some thought it fit best in the R&D unit; a few thought it should report to Marketing as a business function because many of the early project managers held MBA degrees. Second, the project management function has often been paired with other disciplines of related scope, most commonly with Portfolio Management, Alliance Management, or Business Operations. Finally, there has been a continuing organizational debate about whether the project management function should be centralized across the organization or whether project managers should report to the project leads or the disciplines they support. I'll share an opinion on these considerations later.

Regardless of organizational inconsistencies, since the early 2010s, the inclusion of project management has become a necessity rather than a luxury for biopharma companies. Typically, a company will start to look for project management support at the point of their first asset entering the clinic. More recently, companies are starting to look for project managers to join the company even as they plan their first IND.

Another area of evolution over the past few decades has been the scope of services that project managers provide. While the core project management function remains consistent (i.e., to convert strategies to action plans and execute those plans to deliver value), the ancillary services that are provided have become more critical to the business. The original project managers from the 1990s were expected to manage individual projects with basic tools such as schedules, budgets, resource plans, etc. Project managers were then relied on as the primary and central source for project information to support portfolio decisions. Owing to their central, independent role on the cross-functional teams, project managers began taking on the responsibility to drive alignment and enhance team effectiveness. More recently, project managers are being asked to support strategy realization and value maximization for the entire enterprise, not just individual projects —hence giving rise to the project management office that we'll discuss in Chapter 4.

In addition, the role of the project manager is evolving to become more specialized in dedicated areas such as Research, Development, CMC, Regulatory, and Commercial, and more recently into functions such as Safety/ Pharmacovigilance, Medical Affairs, and Clinical Pharmacology. We will describe some of these specialized roles in Section 2.4.

As mentioned previously, there has been considerable variation in how the Project Management department is connected to the organization. While companies will have different needs based on their pipeline, outsourcing strategy, size, and geographical footprint, it is useful to think about the best practice for a "typical" mid- to large-sized company. A typical company might have multiple assets across multiple indications in one or more therapeutic areas, it will have dedicated functional support to cover the scientific and operational aspects of the business, and it will be in a competitive space with other companies, putting a premium on efficiency and speed to market. For this type of company, I feel the best organizational construct is to have a vice president (VP) of Project and Portfolio Management reporting to the head of development. I'll explain why.

First, why report to the head of development? The head of development could be a Chief Medical Officer, Chief Development Officer, Chief Scientific Officer, a President of R&D, or other title, but the key concept is that all molecule project work rolls up to this single individual. One of the key roles I've seen project managers play is in filling the gaps in the project leader's skill set (often business savvy, strategic perspective, or people skills). Since the head of development is often the person who assigns the project leader, having direct access to them helps refine the selection and coaching of the project leaders. Why VP level? The department needs to have the same access to the primary decision maker as other functions, so Director level is simply not enough. Why combine the function with portfolio management? The information flow and alignment on strategy between project and portfolio management is key to maximizing value for the company. While portfolio management is charged with identifying and selecting the mix of projects to run, project management is charged with running those projects and ensuring that the correct delivery assumptions are included in the project valuations so that maximal value can be achieved. We'll dive deeper into the interface between project and portfolio management in Chapter 5.

2.4 WHAT DO PROJECT MANAGERS DO IN BIOPHARMA?

Let's take a deeper look at what a project manager does in biopharma. Fundamentally, biopharma project managers convert complex drug development strategies to action plans and then coordinate the execution of those plans to deliver marketable drug products that bring value to the organization. To do this, biopharma project managers interface with several stakeholders across the organization, including members of a product development core team and subteams, what I call the "project operations" functions, and governing bodies. We will discuss the project manager's involvement in the project team in Chapter 3, project operations in Chapter 4, and governing bodies in Chapter 5. For now, let's describe these groups as follows:

- Project teams are responsible for delivering projects
- Project operations functions are responsible for ensuring projects are being delivered in the best possible way
- Governing bodies are responsible for selecting the projects to execute and committing the organization's resources to those projects

As mentioned previously, the application of the project management discipline has expanded beyond the original assignment to clinical development projects.

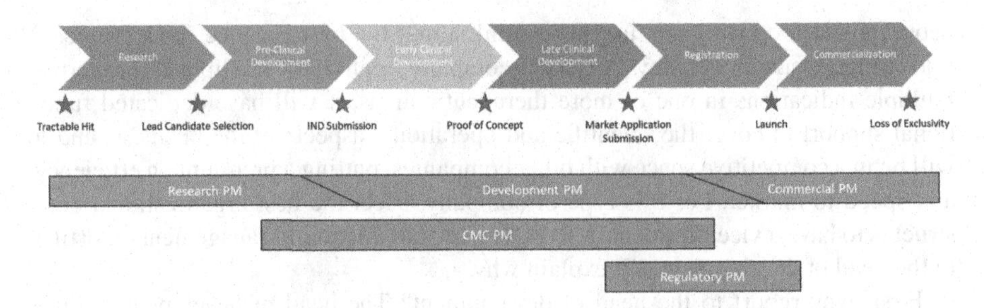

FIGURE 2.4 Project Manager roles juxtaposed with the drug development lifecycle.

Nowadays, many companies have project managers for all stages of development (i.e., preclinical, early development, late development, and commercial) and for many functional lines (e.g., CMC, Regulatory). The sections below provide a general description of the responsibilities of project managers in biopharma as well as specialized project manager roles that are commonly found in mid-sized and large companies (Figure 2.4).

2.4.1 Research Project Manager

Research project managers are focused primarily on activities within drug discovery and preclinical development that culminate in an investigational new drug (IND) application. They are typically assigned when a lead compound has been identified, and they support the project through lead compound optimization and preclinical development. After a molecule enters clinical development, the research project manager will continue to work on the asset team to coordinate any remaining nonclinical work that is needed to support clinical development and regulatory requirements.

The key responsibility of a research project manager is to build and execute a nonclinical development plan that supports IND clearance and NDA submission. While the asset is in drug discovery and preclinical research, the research project manager has sole accountability for managing the asset. When the asset enters clinical development, the research project manager interfaces with the development project manager and CMC project manager to align the nonclinical plan with other development plans (e.g., the clinical development plan, clinical pharmacology plan, and CMC plan).

2.4.2 Development Project Manager

Development project managers focus primarily on activities within early and late development that culminate in a market application submission (e.g., New Drug Application (NDA) or Biologic License Application (BLA) in the US, Market Authorization Application (MAA) in the EU). They are typically assigned just before an IND/Clinical Trial Application (CTA) is filed and while the first-in-human study is being planned. Some organizations will divide the responsibilities of development project managers even further into early development project managers who cover activities leading up to proof of concept and late development project managers who cover activities after proof of concept that lead up to a market application submission.

A key responsibility for the development project manager is to build and execute an Asset Development Plan and Clinical Development Plan that supports a market application submission. The development project manager interfaces with research project managers, CMC project managers, and regulatory project managers to maintain alignment between plans. The development project manager also partners with the development project leader to manage the product development core team.

The generic job description for a development project manager is available in Appendix VI. The job description provides more insight into what a development project manager's responsibilities are, as well as serving as a starting point for a Director-level job description that can be used for hiring.

2.4.3 Commercial Project Manager

Commercial project managers focus primarily on drug product launch and commercialization. They are typically assigned just before a market application is submitted, and they support the project through launch and all stages of the commercial lifecycle (i.e., growth, maturity, and decline).

The key responsibility of a commercial project manager is to create and execute a launch and commercialization plan. The commercial project manager interfaces with the development project manager and the regulatory project manager to maintain alignment between plans.

2.4.4 CMC Project Manager

CMC project managers focus primarily on activities related to drug substance, drug product, and analytical method development. The CMC project manager also coordinates the associated quality assurance and regulatory compliance activities needed to demonstrate that the manufacturing process is controlled and reproducible. To that end, CMC project managers work with the regulatory team to develop timelines to support CMC submissions to health authorities. A CMC project manager is typically assigned during the research phase of development, and he/she supports the asset throughout its lifecycle.

During clinical development, the CMC team supports the core team by ensuring continuous supply of clinical trial material, including the manufacture, testing, packaging, and labeling of clinical trial supplies that will be sent to investigative sites. This requires close integration with the clinical supply team, who is responsible for converting study enrollment projections into drug supply–demand curves. Additionally, because most small and medium-sized biopharma companies outsource manufacturing activities to a contract development and manufacturing organization (CDMO), the CMC project manager is responsible for interfacing with and managing the flow of information with external partners, often in geographically dispersed regions (Chapter 7).

2.4.5 Regulatory Project Manager

Regulatory project managers partner with the regulatory lead, the development project manager, and the regulatory subteam to ensure regulatory activities are executed as planned. These regulatory activities include the following:

- Partner with the regulatory lead to development of the regulatory plan that defines the path to approval that aligns with the asset development plan and clinical development plan
- Triage, review, and QC of documents submitted to health authorities, including INDs/CTAs, NDAs/MAAs, investigator brochures, annual reports, development safety update reports (DSURs), and meeting materials. Tracking of documents submitted to health authorities across the globe and coordination of responses to health authorities' queries
- Monitoring the regulatory landscape to identify risks to and opportunities for the development plan

In addition to the broader support roles listed above, the regulatory project manager also has specific accountabilities for the planning and execution of NDA submissions. Therefore, a regulatory project manager should have expertise in the following:

- Managing cross-functional module meetings and submission team meetings.
- Creating and managing cross-functional timelines for submissions with consideration of key interdependencies.
- Standardizing the best practices for the management of regulatory submissions to ensure consistency across functions, e.g., defining the submission process, creating templates, implementing tools, and building timelines.
- Overseeing the efforts for the preparation and publishing of regulatory submissions in accordance with the submission timeline.
- Managing the timely delivery of a compliant NDA submission.

2.4.6 Emerging Roles for Project Managers in Drug Development

In recent years, the trend toward more specialized project management roles has led to project managers supporting functions not listed above. For example, I have seen dedicated project managers in medical affairs, pharmacovigilance, supply chain, sales operations, and other functions.

2.5 WHO BECOMES A PROJECT MANAGER IN BIOPHARMA?

Project managers in biopharma seldom sought project management as their first career path. This is likely because there are very few academic offerings that prepare individuals for a career as a drug development project manager. Of the people who do enter directly into project management, most have an education in business administration (e.g., MBAs), where they can use their business backgrounds while learning the drug development piece. This may be changing, as more and more universities are offering programs in life sciences, including a requisite course in project management.

Most often, project managers start their career in a different discipline in a biopharma company and, over time, gravitate to project management because their personality and work habits lend them to being effective at organizing work and coordinating teams. Project managers may come from all areas of the business, from benchtop scientists to business support. Indeed, the best project management departments I've seen have a mix of representatives with both scientific and business backgrounds because a successful project management group needs to be able to bridge both areas.

A common path to project management is when a scientist in research or CMC/technical operations wishes to move from a technical role to a business support role. These transitions are most successful for research project managers and CMC project managers, respectively, where deep technical knowledge is needed to successfully build and execute project plans. Another common route to project management is when a Clinical Trial Manager from clinical operations wishes to broaden his or her development experience by becoming a development project manager. Given that clinical trials are complex projects in and of themselves, the skills learned while managing studies is easily translated to managing broader clinical development plans and asset development plans. Based on personal experience, a candidate having project management experience in another industry is a better predictor of success than having technical experience in the biopharma industry. The common belief in the converse can be a barrier to finding great candidates.

2.6 WHAT MAKES FOR A SUCCESSFUL DEVELOPMENT PROJECT MANAGER?

Though the path to project management may differ, it is a person's personality and work habits that are most likely to make him or her effective at the essential components of the role: organizing work and coordinating teams. What are those traits and characteristics? There is no predictable formula for making a project manager successful with every project and every project team, but a few qualities that are generally recognized in good project managers may be found in Figure 2.5.

While the above characteristics are fundamental for development project managers, the competencies below really set apart a great development project manager, one that will continue to drive impact in the future of our profession. In fact, the PMI has recently updated the Talent Triangle to emphasize the evolution of the project manager role from technical experts to project professionals with multifaceted capabilities. This highlights the fact that our profession is constantly changing and evolving, and as new expectations, opportunities, and technologies arise, we need to be prepared for the future of our role (Figure 2.6).

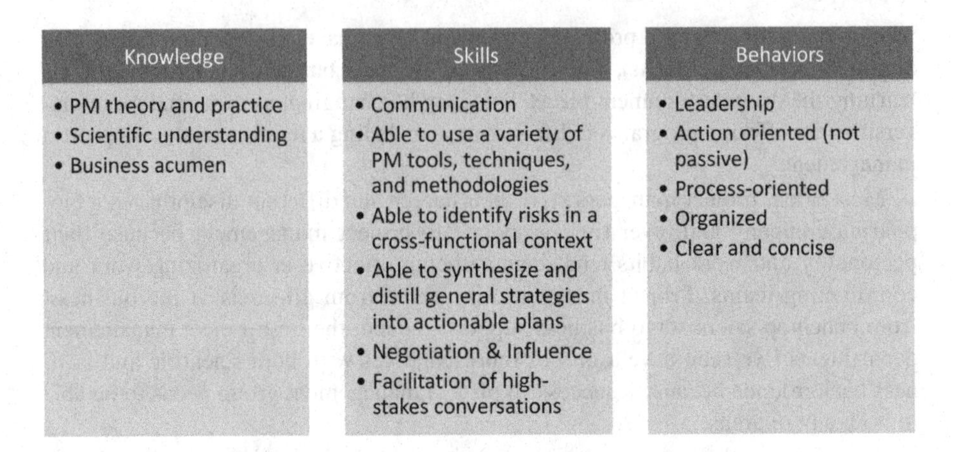

Knowledge	Skills	Behaviors
• PM theory and practice • Scientific understanding • Business acumen	• Communication • Able to use a variety of PM tools, techniques, and methodologies • Able to identify risks in a cross-functional context • Able to synthesize and distill general strategies into actionable plans • Negotiation & Influence • Facilitation of high-stakes conversations	• Leadership • Action oriented (not passive) • Process-oriented • Organized • Clear and concise

FIGURE 2.5 Characteristics of a successful Development Project Manager, as grouped by the knowledge, skills, and behaviors a person must have to be successful in the role.

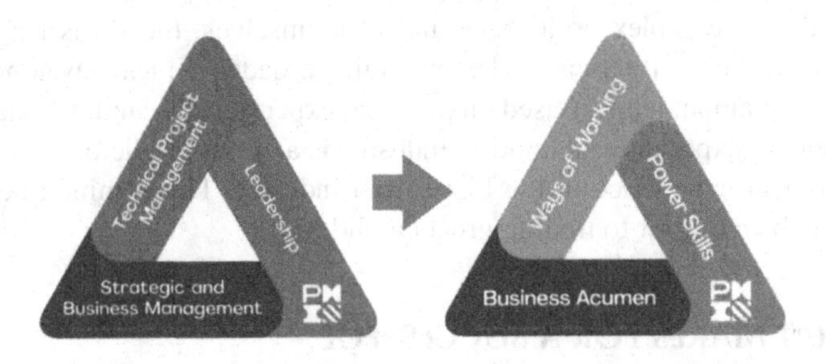

FIGURE 2.6 The evolution of the PMI Talent Triangle emphasizing the transition of the Project Manager's role from a tactical to strategic.

The PMI describes these capabilities as follows:

- **Ways of working**: Whether it's predictive, agile, design thinking, or new practices still to be developed, it's clear that there is more than one way that work gets done today. That's why we encourage professionals to master as many ways of working as they can – so they can apply the right technique at the right time, delivering winning results.
- **Power skills**: These interpersonal skills include collaborative leadership, communication, an innovative mindset, for-purpose orientation, and empathy. Ensuring teams have these skills allows them to maintain influence with a variety of stakeholders – a critical component for making change.
- **Business acumen**: Professionals with business acumen understand the macro- and micro-influences in their organization and industry and have the function-specific or domain-specific knowledge to make good decisions.

Professionals at all levels need to be able to cultivate effective decision making and understand how their projects align with the big picture of broader organizational strategy and global trends.

For a development project manager, the competencies described below align with the evolving PMI framework. I think of these as competencies that will "future-proof" a development project manager, meaning that they will always be in favor regardless of the changes to the profession that come in the future.

2.6.1 Collaboration of Multiple Disciplines

The future-proof development project manager will be able to coalesce a group of individuals with various expertise and experience into a focused team that delivers. To do this, the development project manager should themselves have T-shaped experience and expertise. The development project manager will have a shallow understanding of a broad range of disciplines and a deep understanding of the standard and modern project management competencies. The shallow understandings allow a development project manager to identify when there are touchpoints and dependencies between functional activities and to connect the appropriate team representatives when needed to streamline work, get ahead of potential conflicts, and avoid redundancies. The deep knowledge, skills, and behaviors in the project management discipline will continue to be valued by the project team because it relieves the rest of the team members from having to worry about such things (Figure 2.7).

The activities needed to develop a drug need coordination between many internal functional groups, such as R&D, regulatory, legal, finance, supply chain, sales, and marketing, as well as external partners. The efficiency in coordination between different functional groups is crucial.

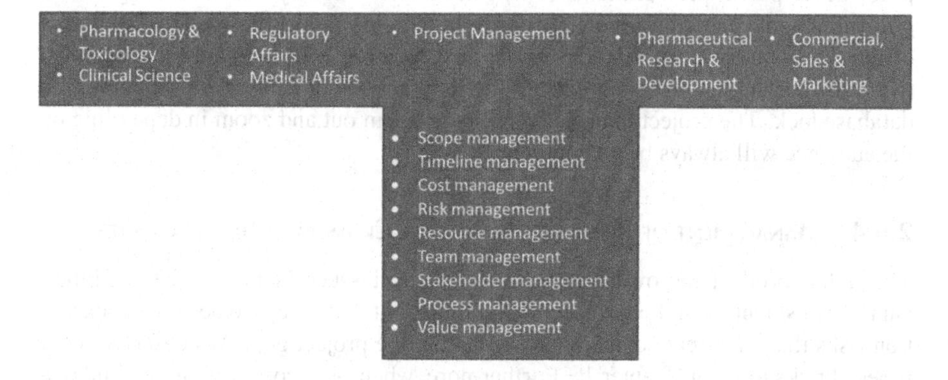

FIGURE 2.7 The T-shaped experiences and expertise a Project Manager should have to be able to collaborate multiple disciplines effectively.

2.6.2 Coordination of Business Processes

Project teams will rely on the development project manager to navigate them through organizational business processes such as obtaining budget and resources, making decisions, addressing issues that require changes to the project plan, and maintaining an environment where the team can perform at its best. In addition to internal business processes, the development project manager must also help the team to work through dynamic external factors such as changes in regulatory, quality, and safety requirements. Therefore, integration of business processes into the execution of projects is inevitable, and seamless coordination of these business processes will enable the team to move quickly and avoid swirl.

The future-proof development project manager will know how best to maximize the business processes to keep the project moving forward. With a good PMO that provides guidance in the form of a PM Playbook or best practice knowledgebase and with access to appropriate records and systems that the development project manager can use to know when change thresholds are being hit, the development project manager is set up for success. Without these components, the development project manager will be left to his/her own resources to learn about and navigate the appropriate business processes.

2.6.3 Communication across Multiple Stakeholders

Similar to being able to collaborate with multiple disciplines, the future-proof development project manager will need to be able to communicate across a broad set of stakeholders. The key stakeholders are the project team, governing bodies, and functional area heads. Additional stakeholders come into and out of the development project manager's hegemony as the project moves through certain stages of development. There may be consultants, contractors, vendors, strategic partners, and regulators that need project information. The development project manager will need to know the right information to share with each of these stakeholders and how to craft the information into a message that the audience can understand.

To be a good communicator across the diverse set of stakeholders, the future-proof development project manager will need to be "zoomy". In a meeting with a governing body, the project manager can describe the value of the asset in the context of the development portfolio; and in the next meeting with the study management team, the project manager can discuss the implications of delayed data cleaning on database lock. The project manager's ability to zoom out and zoom in depending on the audience will always be an appreciated skill set.

2.6.4 Management of Risks and Issues in a Cross-Functional Context

The path to product approval is paved with risks and issues. Not only are there inherent risks of scientific and pharmaceutical failure, but there are also design and execution risks that can create obstacles to achieving the project goal. We describe these types of risks more in Chapter 16. Furthermore, when risks convert to issues and resolution is needed, changes to the project plan that the development project manager has so fervently trained the project team in are inevitable. We discuss these issues more in Chapter 17. The lack of effective risk and issue management can result in delays in the project and decreased team morale.

The future-proof development project manager will be able to manage cross-functional risks and issues with agility and determination. A Risk Management Plan will help to inform the team of how risks will be identified, tracked, and responded to, but an RMP only describes on paper what needs to happen. The development project manager will also need to create an environment where the project team members feel comfortable sharing risks. The development project manager will need to know when to escalate risks for broader awareness and when and how to escalate issues for appropriate resolution.

2.7 SUMMARY

The project management discipline has become integral to drug development, and project managers are now revered as key players in managing the complex projects required to provide evidence of safe and effective drugs. The role has expanded within biopharma beyond clinical development, with project managers now supporting projects in research, CMC, commercial, regulatory, and more. The responsibilities have also expanded from the early days; now project managers use their business acumen and power skills in addition to their technical ways of working to move projects forward. As the responsibilities continue to evolve, it will be important for project managers to future-proof their skill set by being experts at collaboration across multiple disciplines, coordination of complex business processes, communication across stakeholders at multiple levels of the organization, and management of risks and issues in a cross-functional environment.

3 Project Teams

Joseph P. Stalder
Groundswell Pharma Consulting

CONTENTS

DOI: 10.1201/9781003226857-4

3.1 INTRODUCTION

As described in Chapter 2, one of the key responsibilities of the project manager is to manage team dynamics. When team dynamics are managed well, the team can perform at its highest level, resulting in fast decisions and high productivity. To manage a team effectively, project managers need to know the various ways a team can be structured, the roles of each of the team members, and how to create an environment where the team can operate at its fullest potential.

In this section, we will review a few representative team models for common corporate and project situations. We will describe the role of the project manager and project leader on the project team and the roles of some of the essential team members (aka the core team). Finally, we will explore common operating norms and ways of working and propose ways the project manager can manage a team through times of disruption.

First, let's define the "team" that the development project manager is responsible for managing. While there are several names for development project teams in biopharma (e.g., Product Development Team (PDT), Lifecycle Team (LCT), Asset Development Team (ADT), just to name a few), it's probably more useful to define the team in terms of the purpose, scope, and remit. Thus, for this chapter and the remainder of the book, we will use the term *product development core team*, or just *core team*, to mean the group of individuals whose purpose is to maximize the value of a given asset, covering all stages and functions of drug development, and who has authority to make decisions within the parameters of the Asset Development Plan (ADP).

3.2 PROJECT TEAM MODELS

Drug development is a complex and risky endeavor, and, in order to navigate the complexity, specialists covering a broad range of disciplines need to come together to get work done. A fully staffed drug development team may involve 50–75 members. With such large teams, it becomes necessary to break the team into manageable components – hence the *team* becomes a "team of teams". How these teams are organized to make decisions and get work done becomes the project team model.

It is up to the project manager and project leader to shape the team in a way that best suits the project's needs. This refers to both the team structure (the core team, sub-teams, task forces, and meetings) and the membership (the functional and sub-team representation on each team component). As a project moves through various stages of development and faces different challenges, the project team structure and membership may need to change, and it is up to the project manager and project leader to recognize and evolve the team accordingly.

Before we get into the various types of project team models, it is important for project managers to know the typical environment in which project teams operate. Almost all biopharma companies are organized in a weak project management structure, meaning the team members have solid-line reporting to a function and dotted-line reporting to the project. In this type of matrix structure, functional area heads (FAHs) serve as people managers that supply resources to projects and help define their priorities – sometimes in conflict with the priority set by the project manager or project leader. This type of organizational structure can challenge the project manager in the following ways:

- Team members may be under-committed or under-allocated to the project, leaving the project manager to adjust schedules to accommodate overstretched team member workloads
- Communication of project information to the functional line may be inconsistent with the project manager and project leader's intended message, sometimes resulting in swirl or crises when the FAH intervenes at the last minute
- Team members are assigned to multiple projects and may be pulled in different directions at the same time, resulting in decreased responsiveness that the project manager then must address through repeat follow-ups

To overcome these challenges, project managers need to develop relationships with team members as well as FAHs. They need to build credibility and trust with team members by

- Understanding enough of the technical aspects of the team members' jobs to coordinate activities effectively
- Respecting the team members' roles and recognizing their contributions to the team
- Maintaining a clear focus on the success of the project rather than personal gain or departmental agendas
- Making promises and then delivering on those promises.

Project managers may need to engage with FAHs to express the needs of the project and ensure appropriate resources are assigned to it. They can build effective working relationships with FAHs by

- Clearly expressing the needs of the project in terms of vision and planned activities
- Providing feedback to FAHs on team member performance, thereby helping them to assess development opportunities for their group
- Respecting the role of FAHs in assigning the right resource to meet the needs of the project

Another ramification of the cross-functional team is that project managers need to know whom to involve when resolving issues and ensuring information flows to the appropriate team members. This "business savvy" is difficult to learn through reading or coursework; a person needs to gain experience by being in the business long enough to encounter situations that require cross-functional input.

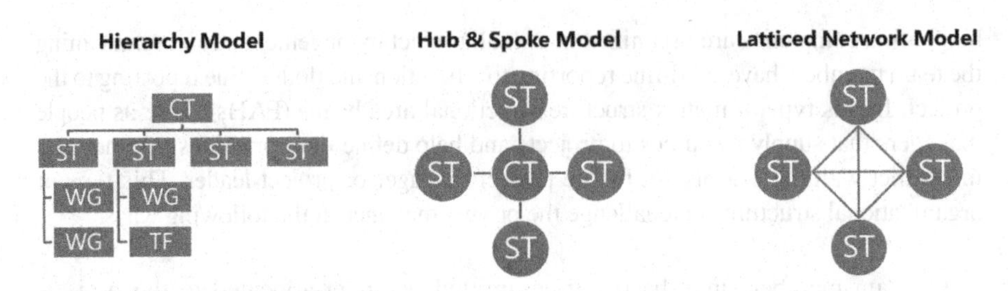

FIGURE 3.1 Visual representation of the common project team models for drug development projects.

With that as a backdrop, let's now look at a few common project team models in biopharma companies. The three most common models are the hierarchy, the hub and spoke, and the latticed network. Each model accommodates connections between teams by assigning a "sub-team", "task force", or "working group" designation. These groups are defined in Section 3.3. The following figure shows a visual representation of the connections between team components and the relative reporting lines (Figure 3.1).

3.2.1 Top-Down Hierarchy

In a hierarchical team structure, the core team is at the top level and sub-teams are at a subordinate level. A further level in the hierarchy may also exist for task forces and working groups, for example study management teams reporting to the clinical sub-team. This type of model is also commonly used to represent global teams, where a global project team is at the top of the hierarchy and regional teams report to it (see Chapter 7).

This model reflects the decision-making authority of the team, where issues escalate up the chain of command based on the relative impact on plans and the breadth of cross-functional implications of the issue. While this model provides strong "command and control" oversight for the project leader, the downside is that issues frequently end up escalating to the core team because the sub-teams do not feel empowered to make decisions. Additionally, to ensure the adequate flow of information to the project leader, more time is spent in meetings that provide updates on the status of activities. While some projects may operate effectively in this way, it is less scalable than the hub-and-spoke model or the network model (Figure 3.2).

3.2.2 Hub and Spoke

The most common project team structure in biopharma these days is the hub-and-spoke model, where a core team serves as the hub and functional sub-teams represent the spokes. This model is meant to de-emphasize the role of the core team as the highest escalation point for decisions and instead empower the sub-teams to resolve issues on their own. The core team, then, serves to coordinate the activities and enable the appropriate flow of information across the team. The core team's decision-making remit, then, is focused on issues that affect the overarching ADP. The core team also serves as a point of escalation for issues that cannot be resolved at the sub-team or when an issue affects more than one sub-team's plans (Figure 3.3).

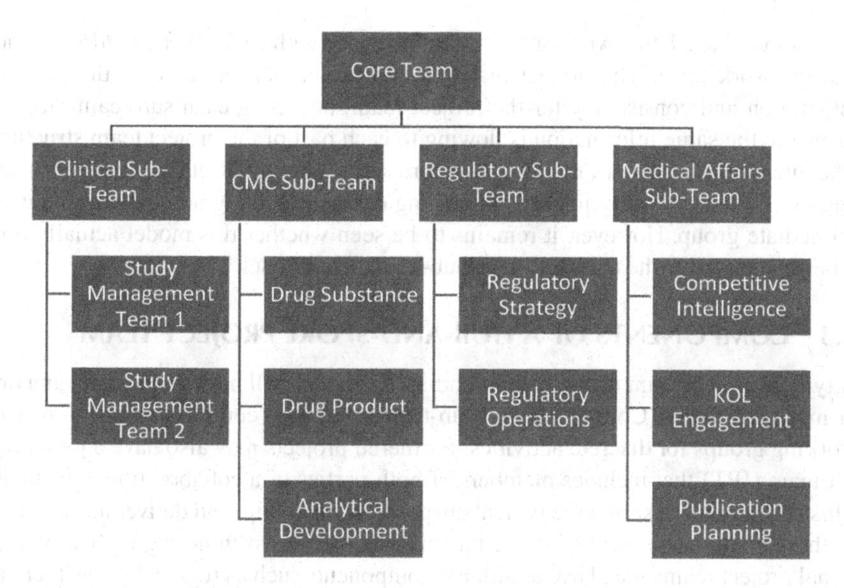

FIGURE 3.2 Example of a hierarchical team structure with sub-teams and task forces reporting up to the core team.

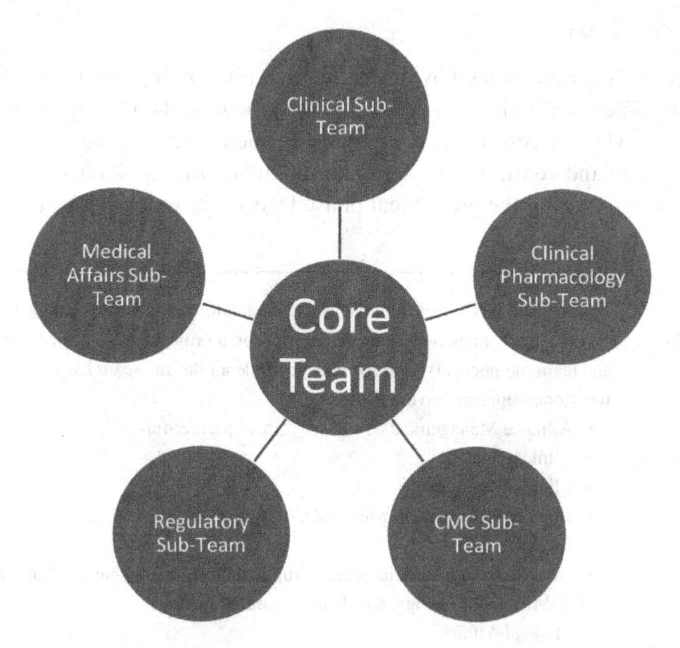

FIGURE 3.3 Example of a hub-and-spoke team structure with sub-teams operating independently and providing inputs into the core team to coordinate efforts across the team.

3.2.3 Latticed Network

More recently, a few biopharma companies are experimenting with a project team model that gets rid of the core team altogether and instead relies on the network of communication established by cross-pollinated sub-teams. This project team model looks more like a latticework of teams that interface with each other to address issues and get work done. The project manager and project leader serve as the point of integration and consistency for the project team, attending each sub-team meeting to ensure the same information is flowing to each part of the project team structure. The intent of this network model is to embrace the agile mindset (see Chapter 6) and empower teams to work quickly by pushing decision-making authority to the most immediate group. However, it remains to be seen whether this model actually provides benefits over the tried-and-true hub-and-spoke model.

3.3 COMPONENTS OF A HUB-AND-SPOKE PROJECT TEAM

A typical project team using the hub-and-spoke model will have a core team and one or more sub-teams. Core teams and sub-teams may also commission task forces or working groups for discrete activities. Partnered projects may also have a joint project team (JPT) that includes members of both parties of a collaboration agreement. This section will describe the typical purpose, membership, and deliverables of each of these components. Out of scope for this section but worth noting is that international project teams may have additional components such as regional project teams, as described in Chapter 7.

3.3.1 CORE TEAM

The core team is commissioned by the R&D governing body to be a cross-functional team responsible for defining the development strategy and ensuring the timely execution of the ADP. A core team is typically formed when a program enters clinical development and continues through market authorization. Some companies may form a core team during the preclinical phase leading up to the IND submission.

Purpose	The core team formulates the strategy and creates the plans for developing a new product.
Membership	The core team comprises an executive sponsor, a project lead, a project manager, and team members. Typical members include all the sub-team leaders (STLs) and functional representatives such as: • Alliance Management (if the program is partner-funded) • Clinical Science • Clinical Operations • Chemistry, Manufacturing, and Controls (CMC) • Commercial • Nonclinical Science (including drug metabolism and pharmacokinetics (DMPK), Toxicology, and Translational Medicine) • Medical Affairs • Regulatory Affairs
Deliverables	• Annual project team goals • Asset Development Plan (ADP) • Target Product Profile (TPP) • Program budget, as part of the Long-Range Plan (LRP) • Commercialization assessment • Product valuation (NPV)

3.3.2 Sub-Team

Sub-teams are commissioned by the core team and serve as the primary working forum to execute specific parts of the strategic plan. A project may have one or more sub-teams, as determined by the project manager and project leader. The sub-team structure varies depending on the needs of the project. Typical sub-teams are the following:

- Clinical
- CMC
- Diagnostic
- Medical Affairs
- Nonclinical
- Regulatory

Purpose Sub-teams provide technical expertise and operational capabilities to the project.

Membership The sub-team comprises an STL and team members. A project manager may be included as needed.

Deliverables

- Development plans (e.g., clinical development plan, nonclinical development plan, regulatory approval strategy)
- Study status and results
- Recommendations on program direction

3.3.3 Task Forces and Working Groups

Task forces and working groups are small, hyper-focused teams that are established to execute a specific activity. Task forces are short-term, temporary constructs, whereas working groups are longer-term constructs. Task forces might be established to handle such activities as protocol development, health authority meeting preparation, or a partnering presentation. Task forces are established by the project leader to conduct a discrete activity and then disband and release resources for other projects.

Working groups, on the other hand, are longer-term constructs that are responsible for specific activity or deliverable. For example, working groups may be established to monitor site activation for the clinical sub-team, assess drug stability data for the CMC sub-team, or keep tabs on an evolving regulatory topic for the regulatory sub-team. Working groups are established by the project leader or STL to conduct a specific activity throughout the project.

Purpose Task forces are commissioned to execute a specific, discrete activity

Membership Membership of task forces is determined by the project leader. Task forces are intended to be lean, only consisting of those members that are absolutely required to complete the activity.

Deliverables

- Work stream proposals and recommendations
- Risk assessments
- Change requests

3.3.4 Joint Teams

Joint teams (also known as joint project teams, or JPTs) may be formed when an organization enters into a business agreement with another company to develop an asset. These teams are usually codified in the collaboration agreement as part of the governance structure that determines how decisions will be made.

Purpose	Joint teams serve as a connection point between the partner organizations. Their purpose is to convey information from one party to another, reach an agreement on priorities, and resolve conflicts that may arise during project execution.
Membership	Joint teams contain members from both sides of the agreement. Typically, an alliance manager, a project manager, and technical representatives comprise the joint team.
Deliverables	• Joint Development Plan • Program updates

3.3.5 Tying It All Together with a Project Team Charter

The word "charter" seems to have a negative connotation in small biopharma companies. Somewhere along the way, team members were dragged through a painful experience of creating a multipage description of every activity the team was expected to do, and then the document was archived and never used again. We, PMs, know that the value of a charter is to help the team understand the purpose, remit, and representation of each component of the project team system. To that effect, the following "mini-charter" template is sometimes all we need (Figure 3.4).

I like to keep a slide deck called "project team operating model" that has a visualization of the project team structure, the team roster, and a "charter" slide for each team component. I can then use the project team operating model slide deck to show changes to the team structure as it evolves throughout development as well as the membership changes as team members cycle in and out of the team. This deck can

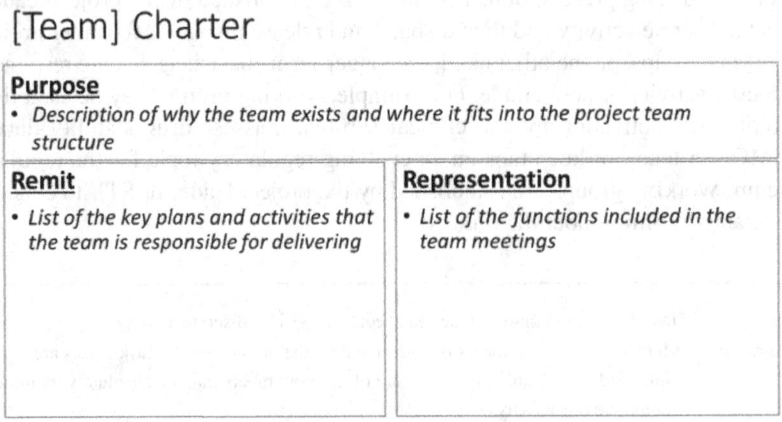

[Team] Charter

Purpose	
• *Description of why the team exists and where it fits into the project team structure*	
Remit	**Representation**
• *List of the key plans and activities that the team is responsible for delivering*	• *List of the functions included in the team meetings*

FIGURE 3.4 Skeleton "mini-charter" document that describes the purpose, remit, and representation of the team component (core or sub-team).

also be used to onboard new members to the team structure so that they're aware of who does what on the project team.

3.4 ROLES WITHIN A PROJECT TEAM

While project team membership evolves as the asset moves through the development stages, the roles usually remain the same. The roles include a project lead, a project manager, STLs, and team members or functional representatives. Tangential to the project team are FAHs and technical experts (TEs) aka subject matter experts (SMEs). In this section, we will describe what each of these roles does on the project team.

3.4.1 Executive Sponsor

The executive sponsor is a member of the company's leadership team who interfaces with the project team to ensure the project team's activities are in alignment with corporate strategies and objectives. In effect, the executive sponsor provides guidance and oversight to the project leader and project manager for issues that challenge the ADP. While the executive sponsor may not attend the core team meetings, the project leader and project manager should have regular contacts with them to maintain alignment and seek guidance. I find a monthly cadence is sufficient to stay aligned because the corporate strategies and objectives usually do not change more frequently than that.

3.4.2 Project Leader

The project leader , sometimes called the Asset Team Leader or Product Team Leader, is often considered the CEO of the asset because he/she serves as the single point of accountability for the development of the asset. In general, the project leader's role is outward facing – managing stakeholders to ensure the project is properly prioritized within the portfolio and adequately resourced to deliver the expected outcomes – and direction setting – defining the vision and strategy for the project.

The project lead is responsible for

- Developing, gaining approval for, and implementing the ADP
- Defining the scope, timing, and priority of activities within the ADP and setting annual project team goals in alignment with the ADP and corporate goals
- Working with FAHs to secure the resources required to deliver the ADP and providing performance assessments for team members to the FAH
- Resolving issues escalated from the sub-teams by making decisions within the parameters of the ADP or escalating issues that need to be resolved at the leadership level, and providing guidance to STLs and functional area representatives (FARs) on issues that can be resolved at the sub-team and functional level
- Adjusting project activities to accommodate budget and resource constraints

- Ensuring that governing bodies, leadership, and other key stakeholders are aware of and agree to changes in project strategy and plans (including recommending termination of the program if warranted).

3.4.3 Project Manager

Where the project leader is the CEO of the asset, the project manager can be viewed as the team's COO. The project manager and project leader form a leadership team within the team, with the project leader setting the project's strategy and direction and the project manager driving the project's execution. While the project leader's role is outward facing, the project manager's role is team facing – managing the team's needs to deliver the project's expected outcomes and creating a high-functioning, highly effective team.

In order to deliver the project, the project manager must coordinate the activities of the sub-teams and functions, often interfacing with STLs and sub-team project managers from those sub-teams and functions to keep plans aligned. I find it useful to have a weekly "leads meeting" or "agenda alignment meeting" with each of the STLs and sub-team project managers to help triage issues and topics across the team system and map out the path for issue resolution. This alignment meeting helps to keep the ADP and sub-team plans synchronized and to recognize and quickly resolve discrepancies between plans.

The core team project manager is responsible for

- Facilitating the development of the ADP
- Facilitating the goal-setting process
- Working with the project leader, team members, STLs, and FAHs to develop timelines and action plans, resource plans, and budgets that support the ADP
- Creating and maintaining the overall integrated project plan and using the project schedule to coordinate activities across the project team system
- Identifying project opportunities, risks, and mitigation strategies in support of effective delivery of the ADP
- Participating on sub-teams and facilitating communication to and from the core team and ensuring relevant issues identified at sub-teams are escalated to the core team
- Collaborating with Finance to maintain an accurate budget and track and communicate key variances
- Collaborating with Portfolio Management to align on cost and timing assumptions
- Providing clear, complete, and accurate project information in reports used to inform the leadership of status and in proposals used for governance interactions
- Ensuring that team meetings are productive and well run, with meaningful agendas and informative minutes.

One of the most important yet intangible roles the project manager plays on the project team is to serve as the "glue" for the team. Arguably, the most valuable project management talent is to address team members' psychological safety, especially where organizational processes fail to adapt to individual needs. Every "high touch" point in a project (i.e., the most critical and sensitive area in a project) can create stress and pressure on individuals, and project managers with high EQ and highly developed soft skills can ameliorate those stresses and provide the needed empathy and "warmth".

3.4.4 Sub-Team Lead

The STL is responsible for the planning and timely execution of activities within the remit of the sub-team. The STL works with the project leader, project manager, FAHs, and TEs to ensure the work plan and design of experiments are scientifically rigorous, within budget, and in alignment with the ADP and core team goals. The STL works with FAHs to ensure the sub-team is appropriately resourced. The STL regularly communicates and aligns operational project strategy and scope with the core team and other key stakeholders carraige return.

STLs carraige return are responsible for

- Defining the activities that the sub-team needs to deliver to meet the ADP and core team goals
- Escalating issues to the core team.

3.4.5 Functional Area Representative

A core or sub-team member represents their functional area (and in some cases, related areas) on the project. The FAR is responsible for generating accurate data and sound recommendations relative to their area of expertise (the departments they represent). The FAR works within their department (and those he/she represents) to ensure project activities are aligned with department priorities, recommendations to and from the team are scrutinized, and key questions and risks are identified. The FAR is accountable to the team for the execution of studies and activities within their area of expertise. The FAR contributes to the development of project vision, strategy, and goals. In working to accomplish team goals, the FAR leverages their solid understanding of the team's results, plans, opportunities, risks, and milestones to effectively collaborate with other team members and TEs.

Common decisions for which the FAR is accountable:

- What risks and issues need to be escalated to the sub- or core team
- Communicates team plans to functional area and ensures alignment.

In practical experience, FARs often do not fully appreciate the extent of their responsibilities in communicating between their team and their functional areas.

Team members may be assigned as core, extended, or ad hoc. Core team members are expected to attend all meetings and contribute to all decisions. Extended members are welcome to attend at their discretion based on the agenda. Ad hoc members are often TEs that are invited on an agenda-driven basis.

3.4.6 Functional Area Head

The FAH is responsible for hiring, managing, training, and assigning resources for within-function and cross-functional work. The FAH appoints STLs and assigns team members to core teams and sub-teams. The FAH assures all functional deliverables are met with appropriate technical and scientific rigor through collaboration and prioritization. The FAH resolves workload conflicts as needed. The FAH oversees the departmental budget, processes and standards, and management practices within his/her function.

3.4.7 Technical Expert

The TE (often referred to as the SME) is responsible for contributing rigorous technical expertise and problem-solving for within-function and cross-functional work and ensuring that deliverables are met. The TE acts as a liaison between the respective function and project team, highlighting workload conflicts, issues, and risks to the FAH and project manager.

Common decisions for which the TE is accountable:

- What study design is most appropriate
- What conclusions can be drawn from experimental data available
- What vendor to use for within-function activity.

3.5 FUNCTIONAL REPRESENTATION WITHIN A PROJECT TEAM

Membership can be categorized as "core" and "extended". Core team members are those who are expected to attend every core team meeting and contribute to decisions being made. Extended team members are those who are optional on the invitation and can attend based on agenda topics for a particular meeting. I have also seen team meeting structures where the core team meets weekly and the focus is on planning and issue management, and the extended team meets monthly where the focus is on updates and alignment. The composition of the core team and the extended will change as the drug candidate advances through the phases of development and different functions are involved more or less.

Often, team members cannot be physically co-located and may even be in different countries, raising the issues of time, geography, and culture (see Chapter 7) (Figure 3.5).

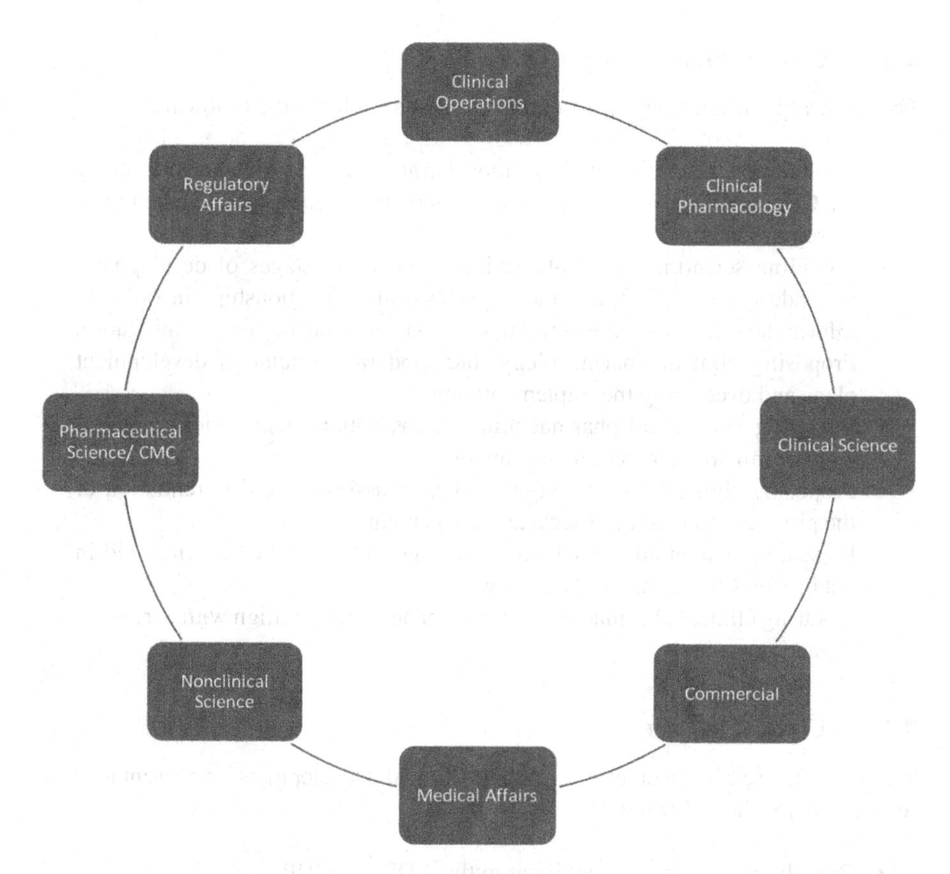

FIGURE 3.5 Example representation of a late development core team.

3.5.1 Clinical Operations

The Clinical Operations representative is responsible for the following:

- Providing operational input to the ADP and Clinical Development Plan (CDP)
- Delivering clinical datasets in accordance with the CDP
- Interfacing with study execution teams to ensure alignment with core team expectations and escalating issues with clinical trial execution to the core team when needed
- Ensuring resources and budget exist to support CDP deliverables
- Ensuring that the design, conduct, analysis, and reporting of studies are performed according to International Conference on Harmonisation (ICH), Good Clinical Practice (GCP), and regulatory requirements.

3.5.2 Clinical Pharmacology

The Clinical Pharmacology representative is responsible for the following:

- Formulating the clinical pharmacology strategy and creating and delivering the Clinical Pharmacology Plan to support the registration and labeling of a drug product
- Providing scientific and strategic input during all stages of development to guide exposure-efficacy and exposure–safety relationships in order to inform the selection of the optimal dose and the frequency of administration
- Proposing clinical pharmacology and modeling/simulation development plans and overseeing the implementation
- Providing nonclinical pharmacokinetic, absorption/ distribution/ metabolism/ elimination, toxicokinetic support
- Proposing clinical pharmacology studies to assist in the differentiation of the product and to aid in lifecycle management
- Presenting data at advisory board meetings and scientific meetings and in publications to extend product knowledge
- Ensuring clinical pharmacology sub-team deliverables align with core team expectations.

3.5.3 Clinical Science

The Clinical Science (sometimes called Clinical Development) representative is responsible for the following:

- Providing strategic medical input to the ADP and CDP
- Providing medical and clinical expertise for all aspects of the development program
- Engaging and interacting with KOLs to obtain feedback on current treatment guidelines, study design, execution, and strategy
- Maintaining current medical knowledge through interactions with scientific advisory boards, external scientific advisors, and key opinion leaders
- Identifying and mitigating risk assessment to ensure appropriate design of clinical studies
- Ensuring medical sub-team deliverables align with core team expectations.

3.5.4 Commercial

The Commercial representative is responsible for the following:

- Developing and describing the commercialization strategy for the product
- Preparing the business case for the CDP
- Providing appropriate commercial input on certain aspects of clinical trial design

- Providing financial reasoning to help priority setting and resource allocation
- Sharing market research findings to further structure the development program for lifecycle management
- Ensuring commercial sub-team deliverables align with core team expectations

3.5.5 Medical Affairs

The Medical Affairs representative is responsible for the following:

- Providing insight into the evidentiary requirements to demonstrate the clinical and economic value of the product in US (e.g., Academy of Managed Care Pharmacy dossier) and ex-US markets (e.g., pricing/reimbursement approvals and health technology assessment)
- Providing insight into the data needs for medical information, publications, and external scientific affairs
- Coordinating the publication planning process
- Providing expertise on patient-reported outcomes (PRO) and leading the development of PRO instruments and evidence dossier to support labeling claims for PROs
- Coordinating the development of HEOR data to support market access
- Leading the development of AMCP dossier
- Ensuring medical affairs sub-team deliverables align with core team expectations.

3.5.6 Research/Nonclinical Science

The Research or Nonclinical Science representative is responsible for the following:

- Providing scientific expertise with regard to preclinical/clinical efficacy and safety aspects of the program and conveying rationale for exploring new indications
- Proposing preclinical and clinical studies to support the mechanism of action, assist in the differentiation of the product, and aid in lifecycle management
- Identifying early signals for efficacy and safety that can be used in go/no-go decision making
- Ensuring all necessary preclinical pharmacology and toxicology studies are completed to support IND/NDA submissions, including authoring of required CTD module sections and responses to health authority questions (pre- and post-approval)
- Liaising with the appropriate groups to address any specific issues related to impurities and/or drug metabolism
- Supporting publications in collaboration with medical affairs
- Ensuring nonclinical sub-team deliverables align with core team expectations.

3.5.7 Pharmaceutical Science/CMC/ Technical Operations

The Pharmaceutical Science/CMC/ Technical Operations representative is responsible for the following:

- Proposing formulation strategies and identifying risks and issues to clinical and commercial supply
- Forecasting timelines, requirements, and risks for CMC development, commercial supply chains, and production
- Forecasting active pharmaceutical ingredient requirements, economical batch sizes, and appropriate marketing presentations (bottles, blisters, inhalers, etc.)
- Delivering drug substances, clinical/study supplies, and suitable drug product formulations and pilot manufacturing processes
- Delivering on the execution of technology transfer leading to commercialization
- Identifying opportunities for intellectual property claims in chemical composition and the manufacturing process
- Ensuring CMC sub-team deliverables align with core team expectations.

3.5.8 Regulatory Affairs

The Regulatory Affairs representative is responsible for the following:

- Providing strategic regulatory direction in all stages of the product development lifecycle
- Providing insight into the current regulatory environment as it relates to the strategic direction of the program
- Assuring timely feedback from global health authorities
- Ensuring the proper staff exists to support submission deliverables in accordance with the regulatory strategic plan
- Ensuring quality submissions including routine and non-routine submissions
- Ensuring regulatory sub-team deliverables align with core team expectations.

3.5.9 Other Key Interfaces with the Project Team

In addition to the above common core team representatives, there may be projects or times in a project's lifecycle when other functional representatives are needed on the core team.

- **Alliance Management**: For projects with collaboration partnerships, an alliance manager may be included on the core team to ensure the ADP and CDP align with the business and contractual aspects of the partnership. The alliance manager interfaces between the core team and the partner team by managing the joint team meetings and serving as the single point of contact for project-related communications. Note that, at companies which have not established a formal Alliance Management function, the project manager often takes on this role.

- **Finance**: The Finance representative is responsible for aligning the annual budget to the ADP and CDP, issuing budget change requests as project plans change, and flagging budget vs. actual variances when they reach managerial thresholds. If the Finance representative is not included in standing core team membership, the project manager should arrange separate recurring meetings to align project and budget assumptions.
- **Portfolio Management**: As discussed in Chapter 5, the Portfolio Management representative is responsible for providing governance committees with the information needed to make strategic portfolio decisions. Typically, a Portfolio Management representative is not part of the core team, so the project manager should arrange separate recurring meetings to align project and portfolio assumptions.
- **IP/ Legal**: The IP/Legal representative is responsible for ensuring the patent family adequately protects the intellectual property of the project. The IP/ Legal representative can provide the project team with market exclusivity timing for the asset. Typically, an annual check-in on the patent situation is sufficient to maintain alignment.

3.6 TEAM CULTURE, DYNAMICS, AND LEADERSHIP

As the central point of contact for the project team, project managers help create a desirable team culture. This often-overlooked facet of the team is critical, and unfortunately, it is usually only given proper attention when there is something going wrong with the team dynamic that results in slippage of project delivery. The best project managers are able to create a high-performance team by creating an environment where all team members can maximally contribute to their capabilities. All teams will develop a team culture – the point here is to actively shape that culture rather than just let it happen.

3.6.1 Characteristics of a High-Performance Team

According to Korn Ferry's *For Your Improvement* (5/e), high-performance teams have four characteristics:

1. **Shared mindset**: They have a common vision. Everyone knows the goals and measures.
2. **Trust**: They know you will cover them if they are having difficulty. They know you will help even though it may mean more work for you. They know you will be honest with them. They know you will bring problems to them directly and won't go behind their backs.
3. **Collective talent**: While not any one member may have it all, collectively they have every task covered.
4. **Team skills**: They know how to operate efficiently and effectively. They run effective meetings. They have efficient ways to communicate. They have ways to deal with internal conflict.

To foster an environment where team members feel comfortable and can deliver to their best ability, the project manager should regularly check in with team members to see how they think the team is operating. I like to administer an anonymous health-check survey twice a year to gauge the team's sentiment and identify issues and trends that can be addressed and confirm that changes that were made in response to previous health checks have had a positive impact (APPENDIX I). In addition, I schedule team-building events and personality assessments every year or so to refresh people's consciousness of team dynamics.

3.6.2 Meeting Operating Norms

It is up to the project manager and project leader to create an environment where project team members can thrive and be successful in spite of project and functional challenges. Most often, this culture is reflected in the interactions at meetings, where the full spectrum of personalities and perspectives are present. Just as the project team needs to know how to operate together efficiently and effectively, project team meetings require some ground rules in order to be efficient and effective. In an article published by HBR (8 Ground Rules for Great Meetings (hbr.org)), Roger Schwarz identifies eight behavioral ground rules for effective meetings:

1. **State views and ask genuine questions:** This enables the team to shift from monologues and arguments to a conversation in which members can understand everyone's point of view and be curious about the differences in their views.
2. **Share all relevant information:** This enables the team to develop a comprehensive, common set of information with which to solve problems and make decisions.
3. **Use specific examples and agree on what important words mean:** This ensures that all team members are using the same words to mean the same thing.
4. **Explain reasoning and intent:** This enables members to understand how others reached their conclusions and see where team members' reasoning differs.
5. **Focus on interests, not positions:** By moving from arguing about solutions to identifying needs that must be met in order to solve a problem, you reduce unproductive conflict and increase your ability to develop solutions that the full team is committed to.
6. **Test assumptions and inferences:** This ensures that the team is making decisions with valid information rather than with members' private stories about what other team members believe and what their motives are.
7. **Jointly design next steps:** This ensures that everyone is committed to moving forward together as a team.
8. **Discuss undiscussable issues:** This ensures that the team addresses the important but undiscussed issues that are hindering its results and that can only be resolved in a team meeting.

Meeting operating norms (sometimes called team guidelines) are best developed jointly by the team under the leadership of the project manager, documented and distributed to the team. They should be developed shortly after the team is formed and should be reviewed and updated any time the team composition changes. This way they become "our" team norms, not "the project manager's" team norms.

3.6.3 Managing through Disruption

Project managers are often the source of continuity for the project team, commonly staying with the project through all kinds of changes in order to keep the project moving forward. As such, there is an implicit responsibility for the project manager to manage the team through times of disruption. Several factors can disrupt the project team's purpose-driven focus:

- Organizational shifts in priority
- Departmental reorganization
- Mergers and acquisitions
- Team member turnover or nonperforming members
- Abrupt handover of responsibility to a new project manager.

All of these factors can destabilize and hinder an effective project team.

3.6.3.1 Organizational Shifts in Priority

It may well be that the legitimacy of the project will be brought into question during a portfolio review and that the project will close. But if the project is to continue, senior management must take great care to ensure that such changes do not weaken the effectiveness of the project team and should devote special attention to stabilizing the efforts of the project manager and the project team.

3.6.3.2 Departmental Reorganizations

Departmental reorganizations, even when project team membership is unaffected, can be disruptive because the affected team members must become familiar with new managers. Affected team members often become distracted by having to create a new working relationship with their new manager, and sometimes it can take months to redevelop this new working relationship. The project manager can work through this by also establishing a relationship with the new FAH and creating a channel for communicating the need to support a team member if issues arise.

3.6.3.3 Mergers & Acquisitions

Mergers and acquisitions often disrupt a project team if there are concerns about rationalizing and downsizing the functional structure. Regardless of the actual team turnover, the concern of downsizing is often enough to distract team members from doing project work. There is no simple fix that a project manager can apply to this problem. Having been through several profound organizational changes, my personal reflection is to encourage team members to focus on project work and demonstrate value to the organization through productive contributions. This will hopefully be recognized when decisions to reduce force are made.

3.6.3.4 Team Member Turnover or Nonperforming Members

Team member turnover is disruptive because, despite the best transition plans, there is always some project knowledge that is lost between the outgoing and incoming member. A change in executive sponsor is disruptive if the new sponsor does not see the project as having the same legitimacy as did the outgoing sponsor. When a new sponsor comes on board, the project manager and project leader should immediately start building the relationship with the new sponsor to make sure the project vision and priority are maintained. A change in team membership is disruptive if ongoing activities and project history are not properly transitioned to the new team member. The project manager can avoid this by ensuring a proper transition plan is created (see the template in Appendix IV), introducing the new team member to the project status, and quickly establishing the new team member's place in the project team.

Project managers may sometimes have to deal with a team member who fails to deliver as expected. Because a project team is only as strong as its weakest player, it is important to provide support to that individual as soon as possible. I like to use the rubric presented in the figure to assess the root cause and apply the appropriate treatment to that individual (Figure 3.6).

3.6.3.5 Abrupt Handover of Responsibility to a New Project Manager

Even a well-planned and time-unconstrained handover between project managers can be somewhat disruptive because team members have to get used to a new management style, but an abrupt handover is far more disruptive because of the loss of project knowledge that occurs. The new PM should do his/her best to use the project management information systems to learn about the project history and status, set up meetings with team members to start building relationships, and quickly re-establish the team culture that supports a high-performance team. For project teams that are already operating efficiently, I've found that wholesale changes to the way

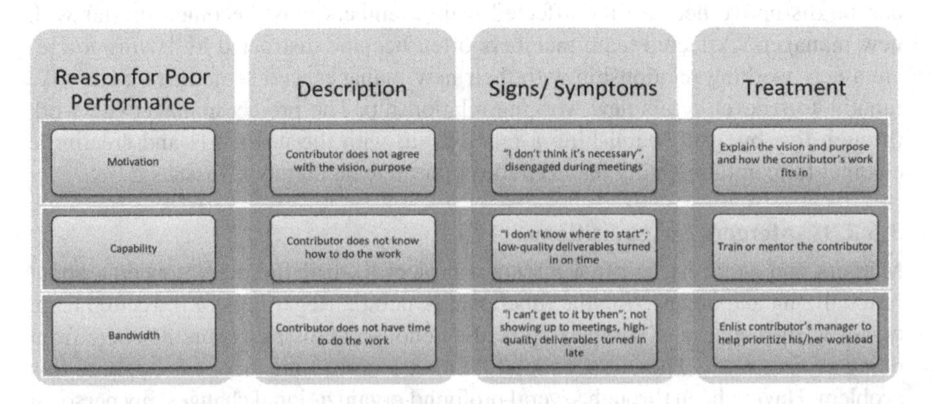

FIGURE 3.6 Rubric for assessing and managing poor performance of an individual team member.

the team and project are managed often alarm and frustrate team members, so a slow and planned introduction to the new project manager's style may be more effective. For teams that are not operating well, sometimes a calculated "refresh and restart" approach is appropriate. The project manager, in conjunction with the project leader, must use his/her judgment to know which approach will work best for the team and the project.

4 The Project Management Office

Norbert Leinfellner
Element Science, Inc.

CONTENTS

4.1 INTRODUCTION

According to PMI's PMBOK (7/e), **project management office** (PMO) is "a management structure that standardizes the project-related governance processes and facilitates the sharing of resources, methodologies, tools, and techniques". The term

DOI: 10.1201/9781003226857-5

"PMO" has become commonplace in business these days, but the exact definition is elusive because PMOs vary in purpose and scope from company to company. I like to define the PMO as being the counterpart to PPM: While a PPM determines "what projects to run" the PMO determines "how projects are run".

PPM: Doing the right things

PMO: Doing things right

Historically, PMOs were viewed as "keepers of the methodology", and many organizations started to view them as bureaucratic institutions that slowed down innovation by requiring too many administrative artifacts. However, more recently, PMOs have adopted a more agile approach, and successful PMOs are viewed as benefiting the organization by increasing the pace and robustness of portfolio decisions. The true value of a PMO, then, is to enhance business agility by supporting project selection and execution through centralized collection and analysis of project information.

In biopharma, the responsibilities of a PMO range widely from company to company. They can exist to support project managers with methodologies, tools, and techniques to do their work effectively; they can serve as the centralized source of project information; they can be responsible for the direct management of business improvement initiatives; or any variation of these and other services. Note the distinction between *projects*, which is the work done to generate revenue for the business, and *initiatives*, which is internal work meant to improve business efficiency. The PMO may support selection and execution of projects and it may be the decision maker on which initiatives to perform.

PMOs have existed in mid- to large-sized biopharma companies for many years, and they are becoming increasingly common in small companies too. As mentioned in Chapter 2, many departments are seeing the value of having a specialized project manager within their group. These project managers usually have a direct reporting line to the department, and a dotted reporting line to the PMO. In this way, the functional area head determines what the project manager will work on, and the PMO guides the project manager on how to do the work. Therefore, it is important for a project manager to understand the various types of PMOs, the various services a PMO can provide, and the responsibility of the project manager to the PMO. This chapter covers these topics, and it ends with some best practice insights on how to establish a PMO in your organization if you are in a position to do so.

4.1.1 What Types of PMO Are There?

PMOs vary in size, formality, maturity, and services. The size and maturity of a PMO is often correlated with the size and maturity of the organization. A small organization with only a few projects will only need a PMO to provide a limited set of services. A large organization with a broad pipeline will need the PMO for more coordination and alignment and a wider set of services.

We can roughly distinguish between PMOs based on their maturity.

Characteristic	Immature PMO	Mature PMO
Processes	Has to create a process for the first time to address a business need	Defined processes that cover a wide range of activities
Systems	Often captures and transfers information manually (and more often paper-based than digitally)	Automated, digital systems to ease the burden on project managers
Services	Building its services and capabilities and attempting to integrate these services into the organization	Well established and deeply rooted in the organization's operational structure

Another way to categorize PMOs is based on the level of control they exert over project managers and the project portfolio. These PMOs can be described as supportive, controlling, or directive.

In a **supportive PMO**, the PMO serves project managers. It provides a menu of tools, templates, frameworks, best practices, and training opportunities from which the project manager can choose to execute his/her project. It exercises little control over how projects are executed, but instead, it creates a community that gives project managers access to the tools and knowledge they need to be effective. A supportive PMO will help to onboard new project managers into the organization (see checklist in Appendix III). The supportive PMO is the least complex and easiest to establish.

Supportive PMOs typically do not have a dedicated head, meaning there is no one purposely hired to run the PMO. Instead, it is run organically by project managers in the organization. In that sense, the supportive PMO can be called a community of practice (CoP) or a best practice forum (BPF).

In a **controlling PMO**, project managers serve the PMO. A controlling PMO creates a methodology that dictates the tools, templates, and frameworks that project managers should use. The goal of this type of PMO is *consistency*. It seeks to standardize the execution of projects to maximize productivity and the intake of project information in order to create consistent reports. Controlling PMOs will help functions to hire new project managers by participating in the interview process, and they provide learning and development opportunities to incumbent project managers. The controlling PMO is moderately complex and usually requires one or more dedicated staff to run its operations.

In a **directive PMO**, project managers report to the PMO. It has the authority to assign project managers to projects. In turn, the directive PMO is accountable for the delivery of projects. Functioning as an organization's centralized authority on projects, a directive PMO brings the highest level of consistency across the pipeline. Directive PMOs hire project managers and are responsible for training the project managers in the organization. The directive PMO is the most complex and often requires several dedicated staff to run its operations (Figure 4.1).

In biopharma, I commonly see small companies have a supportive PMO and large companies have a controlling PMO. As we'll see in Section 4.2, PMOs should

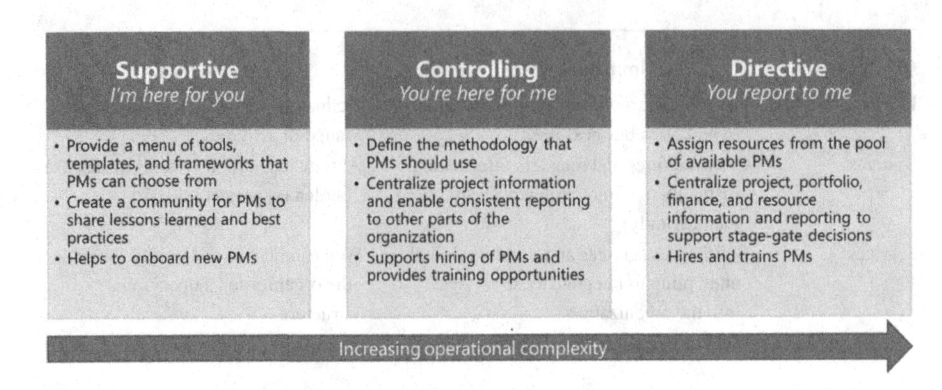

FIGURE 4.1 Common types of Project Management Office

respond to the needs of the customers, and the complexity of the PMO typically increases as the complexity of the pipeline and governance structures within the organization also increases. Ostensibly, this is because more complex organizations require more consistency in order to be productive.

4.1.2 What Does a "Good" PMO Look Like?

Although PMOs vary depending on the needs of the business, there are six imperatives of a mature PMO. Not all organizations will need all six right away, but as a company grows in size and complexity, adding these components will allow the business to remain agile as it scales up. (Figure 4.2).

- **Transparent information**: The PMO should provide project information to relevant stakeholders in a transparent and interpretable way. Transparency is key because the PMO cannot be seen as biased toward one project or another. The full set of information needs to be shared in a neutral and objective manner. The information also needs to be interpretable, meaning the consumer needs to be able to understand the relevance of the information in the context of the rest of the portfolio.
- **Connected business strategy**: The PMO should connect all projects, goals, and efforts to the business strategy so that the organization does not waste

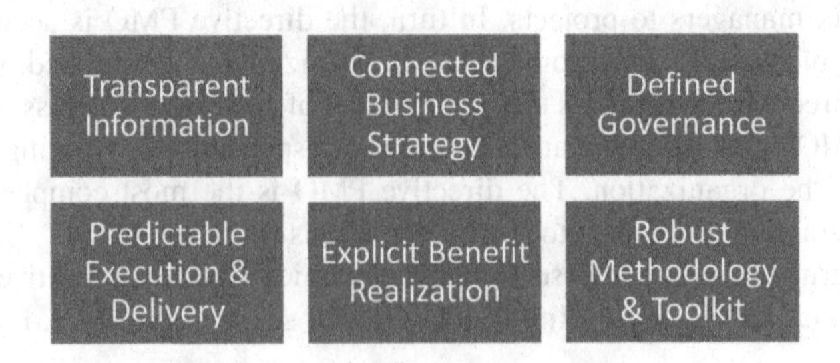

FIGURE 4.2 Key services of a Project Management Office

energy on things that will not drive value for the organization. Assuming the correct business strategy is selected, all work done throughout the organization should be directed toward achieving the outcomes defined by the business strategy.

- **Defined governance**: The PMO should define and facilitate the process for governing the portfolio. This means setting the right questions for the governing body to answer, providing the information needed to make the right decision, and disseminating the decisions to the appropriate stakeholders to follow through. Some organizations have a periodic review of the portfolio, others review each investment opportunity as they come, and the PMO needs a process for conveying the needed information to match the situation for each type.
- **Predictable execution & delivery**: The PMO should enable predictable execution of projects and delivery of outputs. The PMO can track metrics and cycle times so that anomalies can be addressed and future work can be planned more accurately. As part of this imperative, the PMO needs to define standard milestones and cycles that it will track, ideally related to the activities that drive value for the organization as per the business strategy.
- **Explicit benefit realization**: The PMO should be explicit about the benefits that are realized through execution of the portfolio. This often comes in the form of financial results long after the project work is completed, and the product is commercialized for a period of time. The benefit tracking exercise, similar to tracking project metrics and cycle times, can be used to address anomalies so that future benefit predictions can become more accurate.
- **Robust methodology & toolkit**: The PMO should provide a robust methodology and toolkit to allow project managers to do their work efficiently. A clear methodology allows project managers to know what artifacts are needed for each step of the project's life cycle. A complete toolkit allows project managers to navigate through situations quickly by being able to quickly apply the right tool to the situation. The toolkit also includes the project management information system that allows project managers to keep track of activities, budget, and resources in a centralized database that allows summarization and rollup reports (see Chapter 10).

A very mature PMO also drives "continuous improvement" as an ongoing effort to improve all the elements above. It often aligns with the company's desire to provide a constant stream of improvements, and when diligently executed, will have transformational results.

PMOs in companies that excel at continuous improvement start with the belief that success comes from:

- Innovating "how" they do what they do (big and small)
- Engaging all employees (not just project managers) in sharing knowledge and generating improvement ideas
- Exploring better ways to deliver to PMO customers and respond to changes in the internal and external environment

4.2 WHAT DOES A PMO IN BIOPHARMA DO?

The services provided by a PMO depend on the size and complexity of the organization and its pipeline. As a company and its pipeline grow, so does the need for the PMO to expand its capabilities and level of support. This section describes some of the services that I typically see in mid- to large-size biopharma PMOs.

4.2.1 Support the PPM Process

A biopharma PMO can support the PPM process by providing accurate information that is used by a governing body to make decisions about "where to play". As described in Chapter 5, the PMO establishes a method to select and prioritize projects by aligning to business objectives and factoring in other drivers such as regulatory requirements, cost, and development effort, and commitments to other departments like R&D, Sales, Quality, and Manufacturing. In an advanced PMO, the prioritization process is formalized and highly predictable with accurate and precise rules that everyone in the organization is trained on. In a basic PMO, it is often a simple, rank-stacked list of projects or tasks that are solely dependent on resource availability.

PMOs are only as valuable as the information they broker, so the best PMOs have the most up-to-date and complete set of information to enable the business to make decisions. The biopharma PMO should consolidate information from various sources such as Project Mangement, Finance, Portfolio Management, Resource Management, and Process Management to create a dataset and analyses that enable complex decisions. In addition, the PMO should serve as the central repository of ADPs, CDPs, submission plans, timelines, costs, risks, revenue projections, and more.

4.2.2 Support the Project Management Process

A biopharma PMO can support the project management process by developing and maintaining methodologies, playbooks, standards, tools, and templates that streamline and harmonize project management activities. A basic PMO will provide the necessary administrative support to the project manager, but the authority will remain with the project manager on which ones to use. An advanced PMO may have more control over which tools and templates to use.

The PMO is also responsible for selecting and maintaining the project management information system that project managers use to manage their projects. Chapter 10 describes these systems in more detail. It is important for the PMO not only to set up the system but also to make sure project managers know how to use it correctly. So, training and auditing is an important component of the PMO's services that cannot be forgotten.

4.2.3 Support the Annual Planning Process

The PMO is essential to implementing the annual planning process. Not only are project managers in the best position for creating project goals that roll up to corporate goals, but they also provide project demand information that sets the assumptions for budget and resource planning exercises. Since the PMO typically oversees all project assets within the organization, it is a great partner for the Finance department to estimate the annual budget and headcount plans. These projections are fed into the long-range plan (LRP), and they become the baseline by which Finance reports the budget-versus-actual analyses. Chapter 3 provides more details on the interface between project managers, the PMO, and Finance.

The PMO, with its broad, integrated view of the portfolio, can support the resource management and capacity planning process by providing demand information to enable appropriate portfolio decisions. More detail on the resource management service is available in Section 4.2.7.

In a small company with a basic PMO, available resources are often scarce and assigned to multiple projects and tasks. There is no regular or periodic resource planning exercise other than one driven by the annual budget-planning process. An advanced PMO is involved in, if not responsible for, formal and periodic workforce capacity planning that allows the assignment of cross-functional resources to projects by their exact skills and availabilities.

Capacity management and demand management are part of the portfolio management process described in Chapter 5.

4.2.4 Facilitate Governance

In addition to providing information to governing bodies as described in Section 4.2.1, biopharma PMOs also often run the project governance process for their organization. Project governance is a key decision-making process in the organization, and it is best to set it up as a distinct governing body from the one that manages the operation of the business. Whereas operations governance is often run by a Chief of Staff or Chief of Operations, project governance is best served by the PMO because it covers all projects in the portfolio and it has the remit to track project goals for value realization.

The PMO should set up a mechanism by which project teams can request a meeting with the project governing body (or vice versa). The PMO will then schedule the meeting and coordinate the distribution of meeting materials, including the pre-read, presentation materials, and minutes with decisions and actions clearly described. Often, the PMO head will be the one to write the minutes. The PMO also sets up the document repository where meeting materials and decisions are recorded, ensuring that project teams get access to the decisions and actions promptly.

4.2.5 Build Institutional Memory

The PMO is in a unique position to capture, organize, and distribute project artifacts from past projects that offer valuable insights into the planning, conduct, and close-out of similar projects in the future. As described in Chapter 14, this is most effectively done via a knowledge base, an online repository of curated information that can be drawn upon by a user to learn about past successes and failures. In the quest for continuous improvement, the PMO will often establish the knowledge base and then rely on its project managers to keep content updated and relevant.

The PMO should also provide a forum for sharing lessons learned and best practices among project managers across the organization. This can be called by many things (BPF, CoP), and the topics addressed can be wide-ranging and dynamic, responding to problems as they come up, identifying ways to improve, and preparing project managers across the organization to deliver on their projects. This venue can be a place to train project managers on the methodology and tools that are available. It can also be a way to make project managers aware of more general learning & development opportunities such as conferences, books, and white papers. Finally, it is a way for non-project managers in the organization to learn more about project managers, potentially offering a pipeline of new candidates.

4.2.6 Set Common Corporate Culture

The PMO sets common project culture through communication and training on techniques, methodologies, and best practices. Often, it is seen as a reference of how things should work and as a standard of culture that a company wants to establish. One of the most important qualities in a successful corporate culture is transparency where the PMO plays a central role to provide information that is relevant and accurate to support effective decision-making and avoid surprises. Lastly, I have seen PMOs as being a partner in writing and communicating corporate policies, simply because they understand the long-term and short-term needs of the organization.

4.2.7 Support Resource Management

PMOs can support the resource demand and capacity planning activities for the organization. With its overarching view across project priorities, schedules, budgets, and more, the PMO again provides the centralized venue for collating information that can then be used by other parts of the organization for planning purposes (e.g., the annual planning process and the LRP). We will see below that this is often done through dedicated resource managers (RMs) that help with workforce and capacity planning. Appendix V has some term definitions that can help the PMO describe the RM service they will provide to an organization.

The PMO also sets the standards and definitions for resource management. Some considerations include the following:

- whether to use generic FTEs or named individuals
- the time increments (days, weeks, months, and quarters) for timecard reporting and demand estimation
- the scope of time tracking with respect to project vs. non-project (i.e., overhead) work, and which line functions will be included (e.g., trial operations, technical operations, support staff, and commercial staff)
- the definitions of "under allocated" and "over allocated" resources (e.g., if a person is more than 110% allocated, does the line manager need to take action to shift work off that person?)
- the algorithms that are used to predict demand for future work
- the determination of "what's in" and "what's out" of the demand plan based on the organization's long-range plan (LRP)

The standards and definitions that the PMO creates are used by portfolio managers and project managers to plan capacity and allocate resources, respectively. The PMO may also be involved in collated workforce utilization information such as percent allocation and planned vs. actuals and providing this information to governing bodies, Portfolio Management, or Finance.

A word of caution to PMO leaders who wish to establish an RM capability in their organization: The level of detail can quickly become unmanageable for a small PMO, and once the service is put in place, it needs perpetual maintenance to "feed the beast". While the service is certainly useful for many organizations whose pipelines are growing quickly and who need to scale the workforce to support the work, using your project managers to do resource management will take away their time for managing projects and project teams. At some point, you will need to weigh the cost of having dedicated RMs . In my experience, an RM can usually handle about 500 employees, so a company of 1000 employees will need 2 RMs, and so on. In addition, while many PMOs start with a simple spreadsheet to perform resource management, many PMO leaders quickly find that a more robust tool is needed to be able to do the job efficiently (see Chapter 9).

4.3 WHO ARE THE MEMBERS OF A PMO?

The membership of a PMO at a small company typically comprises of a PMO leader and the organization's project managers. In larger organizations, the PMO may also include resource managers, process managers, change managers, communications specialists, or learning & development specialists. This section will describe the responsibilities that the more common roles have with respect to the PMO.

4.3.1 PMO Leader

Depending on the size of the company, a PMO leader can be a dedicated headcount or a part-time role of one of the project managers in the organization. A good PMO leader oversees the team members in the PMO and takes responsibility for the quality and value of each project under his/her care. This involves collaboration with project managers and reporting to the executive staff of the organization.

PMO leaders succeed by facilitating project planning, analyzing financial information, modifying processes, and ensuring proper documentation for the projects they're overseeing. This is done by both focusing on details and by being a "systems thinker" that keeps an eye on the big picture.

Naturally, a PMO leader has to function under pressure, have strong interpersonal skills and be able to juggle different projects.

Detailed responsibilities of the PMO leader include:

- Creating and deploying the methodologies, tools, and templates for project management across the enterprise
- Establishing metrics and standards that will be used to analyze project health, goal attainment, value realization, and opportunities for process optimization
- Creating and deploying standardized reports for conveying project information to stakeholders

- Facilitating executive decision-making around project selection and prioritization
- Establishing processes for tracking resource demand and managing resource constraints across projects
- Establishing processes for centrally tracking and summarizing project risks
- Supporting the hiring of project managers across the organization, offering an assessment of candidates' technical project management skills and experience
- Training new project managers on the PMO's methodology and processes
- Coaching and mentoring project managers in the PMO and across the enterprise

With these responsibilities, a successful PMO leader becomes the person to contact when business leaders need help with project work. The PMO leader is also the person the other project managers go to when they need help executing their projects.

4.3.2 Project Managers

Project managers are the life blood of every PMO. They are responsible for delivering the services set up by the PMO and providing project information to the PMO. Not only are PMs the ones who use the tools, templates, and methodologies that the PMO creates, but also they are working on the front line with project teams, and so they are able to identify what processes and workflows need improvement or standardization.

Project managerss are also responsible for providing project information back to the PMO so that it has complete and up-to-date information to support decision-making. This information can include goal status, schedules, budgets, resource utilization, and risk assessments.

4.3.2.1 Strategic Role

The project manager, along with other high-profile stakeholders, uses methods and processes set out by the PMO to propose and plan a project such that it aligns with the strategic goals of the business unit and company overall. This includes providing information and participating in discussions regarding the project portfolio. In large companies where the product portfolio includes many products that are managed across multiple projects (and hence managed in a multi-project portfolio) it is important to take into account the relationships among projects, the resource pools, and even cross-product dependencies. In close alignment with the project sponsor, project leader and the team members, the project manager also defines and documents the project's measurable success criteria while following the PMO's guidelines for storing and retrieving project metrics. Typical success criteria include:

- **Schedule**: Deliver the project outcome (often a new product, a feature to an existing product, or a strategically important piece of documentation) on time. The delivery date is often defined as a specific day, week of year, or business quarter.

- **Scope**: The number of value-generating features or properties of a product, often documented in a user needs document, or marketing requirements document.
- **Quality**: According to PMI's PMBOK(5/e), quality is "the degree to which a set of inherent characteristics fulfill requirements". The project manager and project management team have a special responsibility to balance quality and grade (a category or rank assigned to products or services having the same functional use but different technical characteristics). In short, the main objective here is to deliver what the customer needs – nothing more, nothing less.
- **Cost**: How much can the project cost, including people cost, expenses, capital expenses, licenses, cost for trial and regulatory submissions, and material cost to put the initial product on the shelf, if applicable.
- **Strategic value**: The strategic value of a project is often determined in the context of project portfolio management.

All the criteria above play an important role in project portfolio management – see Chapter 5. After these success criteria are defined and approved by the project sponsor the project manager creates the project plan.

4.3.2.2 Tactical Role

The more tactical parts of a project manager's role include the day-to-day activities of active project management, the reporting, the documentation of metrics, and the interaction with the PMO. These activities include:

- **Planning and scheduling**: The project manager owns the project plan, which follows the template provided by the PMO. At a minimum the project plan includes: The project description, its success criteria, the structure and specifics of the team, the way the team will implement the quality policy and the way the quality of both the project and the product will be assured during the project, the resources required to deliver the project outcome, and any additional activities necessary to carry out the projects for staying within the success criteria.
- **Reporting**: The project manager reports project status to the project sponsor and senior management, following reporting guidelines provided by the PMO. Reporting can be done periodically, or just in time when escalation is needed. Sometimes the PMO does the reporting in a more centralized manner and the role of the project manager is to provide all the relevant information.
- **Managing project risks**: The PMO typically provides a process for project risk management. The project manager executes this process by conducting initial and periodic project risk assessments and by proposing and implementing mitigations.
- **Communication**: The project manager conducts team and stakeholder meetings. He/she also has the responsibility to disseminate information up and down the organization. The PMO is also regularly informed and given feedback on which services work well and which don't.

4.3.3 Resource Managers

If your organization has resource managers (RMs), it's helpful to include them in the PMO. project managers and RMs have a co-dependent and mutually beneficial relationship. Project managers need human resources to effectively execute project plans, and RMs are responsible for managing the supply and demand of human capital. Project managers provide the demand for project work so that RMs can plan the resource supply (i.e., capacity planning), and RMs in turn ensure that the appropriate resources are available for projects when the project manager needs them. Therefore, it is beneficial to align on how the project planning information gets transferred to the RM group so they can do their job, and the PMO can serve as a venue for establishing ways of working and transparent information flow between the two functions.

4.3.4 Process Managers

Similar to RMs, if your organization has process managers, it is helpful to include them in the PMO. Process managers, sometimes called business process managers or business analysts, are responsible for improving and accelerating business processes. They do this by monitoring cycle times for certain processes and intervening when cycle times start to creep up or if there are outliers to a process that need to be addressed. For example, protocol development typically takes 8 weeks. If the cycle time starts to trend up to 12 weeks, the process manager may take action to evaluate the current state process, identify bottlenecks, determine why the changes are happening, then implement process improvements, if appropriate. Therefore, process managers depend on the project schedule information provided by project managers, and project managers benefit because their projects will run more efficiently through the work of process managers. Again, the PMO is a good venue to establish ways of working and transparent information flow between the two functions.

4.4 WHAT IS THE BEST WAY TO ESTABLISH A PMO?

Several books are available for building and maintaining a PMO, and we will not go into detail here on those topics. However, if you are in the situation of starting a PMO from scratch, here are a few key considerations to keep in mind.

4.4.1 Understand Your Stakeholders and Build the "Why"

Before establishing a PMO, the PMO leader should consider its stakeholders. A PMO's success is measured by the perception of its customers and stakeholders, so identifying the stakeholders' needs, priorities, and motivations at the outset will ensure the PMO delivers services that its customers perceive to be adding value. In biopharma companies, the stakeholders usually are:

- The project sponsor, an individual who is responsible for ensuring appropriate governance mechanisms are in place for the project, and the project is continually measured and monitored.

- The executive leader of a business unit or the organization. This is a senior leader usually in charge of setting the business strategy and often acts as the project sponsor or delegates project sponsorship to another senior leader.
- The product owner or product manager: Those who define the product in terms of user needs, product features, and performance, that is, those who define the "scope" of the project. They typically also calculate the return on investment (ROI) and hence give ranges of the budget, which need to be approved before a project can be started.
- The project sponsor is accountable for the achievement of the project objectives as specified in the business case, and providing senior management level support for the project.
- The quality representative enforces compliance with the quality system as well as internal and external standards, that is, those who define the "quality" boundaries of the project.
- The regulatory specialist defines the regulatory needs to market the product, following local, national, and international laws and guidelines.
- The clinical engineer or clinical specialist who understands the intended use, indications for use, patient needs, and is often tasked to play a role in the clinical testing of the product, on patients or animals. They can have a technical background (such as biomedical engineers) or a clinical education, such as medical doctors.
- The project (design) engineer is tasked to develop the product under tight restrictions of time, budget, scope, and quality. These include scientists, pharmaceutical engineers, chemical engineers, biomedical engineers, technicians, and those in product development that includes devices and hardware and software engineers.
- The manufacturing engineer will oversee defining, establishing, and executing processes of manufacturing the product, usually in a larger team and under the leadership of the Operations department.
- The operations representative is often in charge of Supply Chain, Inventory, Facility Management, and other operational responsibilities, which can include Manufacturing.
- External partners: Other business partners, such as clinical research organizations (CROs) who design and execute clinical trials, external consultants and contractors, and other third-party vendors that have a stake in the project through a contract that defines clear deliverables for them.
- Financial accountants and controllers need to interface with the PMO and are "customers" for their resource plans, the forecasted need, workforce planning, and the calculation of actuals that feeds back into the budgets.
- Finance providers
- Special interest groups
- The PMO members: PMO leader, project managers, RMs, process managers, and other project management associates who are involved in planning, tracking, and managing project work in terms of resources, budget, and optimization (Figure 4.3).

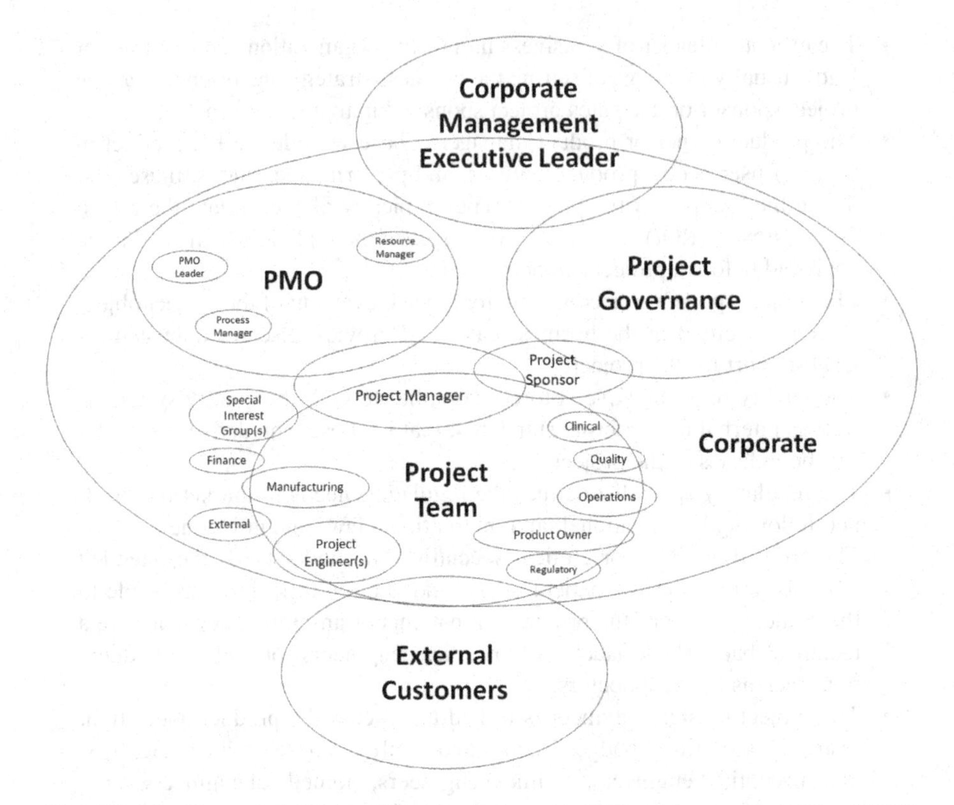

FIGURE 4.3 Key stakeholders and interfaces across the Project Management Office (PMO), Project Team, and Project Governance.

Creating customer value by having a customer-centric mindset

Everything a PMO does and every service it provides should be with that customer in mind. Below are some principles that all PMO Leaders should be aware of to ensure a customer-centric mindset.

Understand what the customer values

Once we know who they are and how they serve, we must ask "why?" Why do the customers need a PMO? We must know what they value enough to ask for or need help and once we ask questions, we should follow with silence. Our objective here is to listen, not talk. How successful have they been in getting their projects delivered with high-impact outcomes? Listen to their story and let them talk about what matters most to them.

Understand the world the customer is in

It's so easy to start building services that we "just know" the customer needs and wants. But do they really want a complete set of templates for them to fill periodically in the project process? The reality is that if our customer is drowning in chaos, they probably cannot even look up long enough to grab a template. We must always meet them where they are and then gently guide them where they need to go. This means that we may have to start by making their lives easier so that they have the time and space to learn something new. The priority is to make it easier, and only later growing capabilities.

Develop a roadmap together

The journey starts with the customer in their current state and is developed together based on pain points they identify and business objectives they must accomplish. It is important to look at the biggest pain points they are experiencing and to evaluate the ones we should solve quickly. This creates momentum and trust. Then, the PMO can establish a thoughtful plan of capabilities that are rolled out over time. It will take them a lot longer to implement new capabilities or engage in new services and see value than it will for the PMO to create them.

Speak their language

As we identify the services that can provide value for the customer, consider how we will talk to them about the capabilities. A project manager needs to speak to someone in the way they need to hear information and help them understand how they can benefit from services. They need to hear about the outcomes that are relevant for them. A skilled project manager is able to flex communication style and can achieve credibility in the process. First the customer will be less interested in the mechanics of the service unless their value can be explained.

Every customer is a partner

Without the buy-in from customers the PMO has no value. It is important to establish a set of guiding principles that both will follow to ensure strategic alignment, transparency, predictability, reliability, and ultimate return on investment (ROI). The PMO must be clear on the importance and their role in providing a stellar customer experience.

4.4.2 Define the Services and Structure

There is no "standard" PMO in life science companies. Each organization will want the PMO to provide different services, such as budgeting, resourcing, and risk analysis, to support the project portfolio. Thus, you should begin your PMO by understanding what the business needs and then build the capabilities and structure to address those needs.

4.4.3 Get Executive Buy-In

Establishing a PMO almost certainly will require some amount of organizational change. Therefore, you will need to gain the necessary support from executives and management in order to be successful. This step is the basis on which a successful PMO is built. Any shift in organizational structure and process must be fully supported by all before any changes can take place.

4.4.4 Identify the Processes and Workflows that Need Standardization

Now that you have your structure and team hammered down, you need to move onto the PMO standards. These standards need to be defined as they will ensure that there is consistency across the organization and all its projects. Being consistent or standardizing your project management process will make the management of your projects easier in the long run.

4.4.5 Train Your Project Managers

Much of the training will focus on the standards set out in the previous step. All individual skill sets should be considered when the training program is being drafted up. The PMO should have up-skilling training programs available in the future as the PMO evolves.

4.4.6 Continue to Examine and Improve Your Process

As discussed earlier, PMOs start small and immature and then grow and mature along the way. A healthy PMO has project managers who want to continuously improve the PMO's services. This means introducing new challenges, fine-tuning existing infrastructure, contributing to the PMOs knowledge base, improving templates and tools, as well as supporting one another with experiences and lessons learned that increase project success.

To understand the areas that need improvement, the PMO leader should identify and measure key performance indicators (KPIs) that reflect the customers' perceptions of success. The KPIs will depend on the services that your PMO provides. Any issues or improvements that are needed can then be highlighted and possible solutions can be examined.

4.5 SUMMARY

The shape and size of the PMO in a given organization largely depends on its business needs. Smaller organizations, like startups, may be well served by having a "one-person PMO" that is executed by a senior project manager. As the business grows in size and maturity, there will likely be multiple project managers at work, along with project portfolio and RMs . This is when the organization needs to establish a PMO that matches the maturity and capacity of those roles and that provides a "methodology" backbone for them. Regardless of the size, the PMO guides the project manager on <u>how</u> to do the work, <u>how</u> projects are run, and <u>how</u> the individuals are engaged. The six imperatives of a "good" PMO are always relevant: transparent information, connected business strategy, defined governance, predictable execution & delivery, explicit benefit realization, and a robust methodology and toolkit. You will not achieve those benefits on day one and building those in time may look different for each organization. It is important to understand the stakeholders and their needs. Each of the stakeholders requires different services and you want to prioritize their provisions carefully. However, every PMO will benefit from the unconditional support and buy-in of their executives to be able to unlock their full potential.

5 Project Portfolio Management

Mark Christopulos
Strategy Execution Consultant

CONTENTS

5.1 INTRODUCTION

As we saw in Chapter 2, project planning information is maintained by the project planning arm of the project operation function, and project planning is supported by a triad of Project Management, Portfolio Management, and Finance. In this chapter, we will focus on the interface between the project manager, portfolio manager, and the Finance department (Figure 5.1).

DOI: 10.1201/9781003226857-6

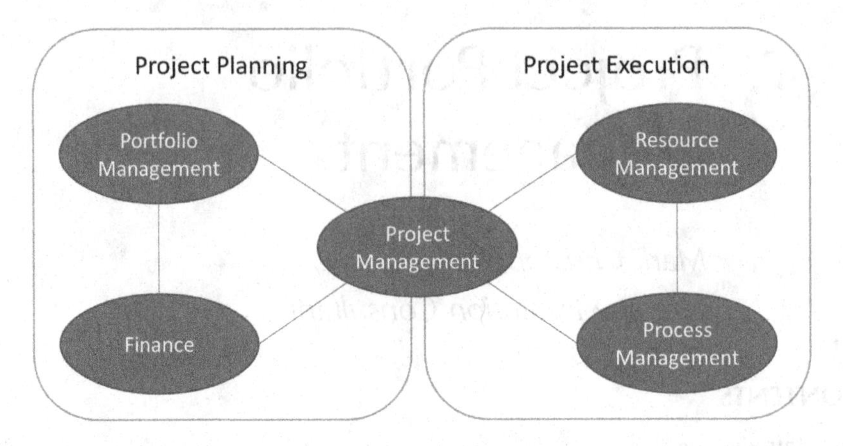

FIGURE 5.1 Conceptual representation of Project Management as the interface of project planning via Project Portfolio Management and project execution via the Project Management Office.

5.2 HOW DOES A PORTFOLIO MANAGER'S ROLE COMPARE TO A PROJECT MANAGER?

The role of the portfolio manager is similar to the role of the project manager, with the major differences being the point of reference. The project manager's role typically aligns to the following:

- seeks to confirm that all the tasks in a project plan are aligned to a clear scope objective,
- works closely with the project sponsor to ensure alignment between project problem statement, project scope, and committed project benefits, and
- tracks project execution to ensure that quality, budget, and schedule constraints are managed.

Whereas the portfolio manager's role, while similar, aligns to the following:

- seeks to ensure all projects in a roadmap are aligned to a clearly defined strategy,
- works closely across project sponsors to ensure alignment between project deliveries and committed portfolio benefits, and
- tracks project execution to ensure that value, budget, and schedule constraints are managed.

Just like the project manager, the portfolio manager naturally also manages risks, issues, resources, constraints, and stakeholders. The key difference for the portfolio manager is that these are all managed at the cross-project level. This means that for a risk or issue to be on the portfolio radar, it must either affect multiple projects or inordinately impact the overall value delivery of the defined portfolio (Figure 5.2).

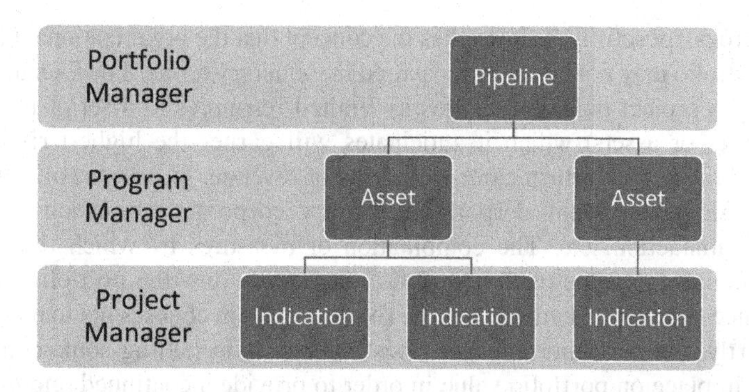

FIGURE 5.2 Typical assignment of remits to Portfolio Manager, Program Manager, and Project Manager.

As an example, the core team might identify a moderate risk associated with the scaling of the underlying technical platform which supports a specific project in the pipeline portfolio. The typical project manager's role would encompass identifying, rating, and prioritizing the risks within the project context. The portfolio manager's role, on the other hand, would involve being aware of this important risk while being clear on how this same risk might also be impacting other projects connected with the same platform technology. The intent here is all about managing overall portfolio value versus diversified portfolio risk – that is, this risk under analysis may affect one particular project which represents only 5% of the overall portfolio value, or that same risk may impact several projects, in which case a much greater amount of portfolio value could be at risk.

In a different example, a core team might also identify a moderate risk involving the underlying platform for a pipeline project, but this time let's say the project represents 30% of the portfolio value or, even worse, let's say that that project is a foundational project that unlocks key capabilities for follow-on projects that represent 60% of the downstream portfolio value. These are both examples of risks that would need to be tracked at the portfolio level and thus necessitate visibility from the executive team.

In numerous places across this chapter, we use the term "risk" or "risk-adjusted return" in order to describe the uncertainty around the value expected from a particular project. While most decision-makers tend to take project deliveries at face value, it is important that the portfolio process account for the potential downsides (and upsides) of delivery uncertainty. A therapy in hand today is worth more than an identical therapy under development for delivery next year. The process, therefore, needs to adjust the stated value of each asset to account for the time, knowns versus unknowns, and possible positive and negative outcomes that could occur prior to delivery.

5.2.1 Portfolio Manager Accountability

Ideally, a portfolio manager is held accountable to delivery of value against the strategic objectives of the enterprise. Although the term "value" can be subjective, it is applied here purposefully, as it invokes the concept that the organizational utility of a project portfolio may not be solely measured in monetary terms. The Executive Team sponsoring a project portfolio applies its limited resources to a set of investments (aka projects or assets) which it anticipates will garner the highest risk-adjusted return over time. That return can be in terms of revenue, gross margin, cost reduction, strategic achievement, corporate capability, corporate acquisitions/divestures, customer satisfaction, etc. The combination of measures by which the executive sponsors define the value of the portfolio is precisely how the portfolio manager's performance should be rated. Even if the Executive Team chooses not to measure this value overtly, the portfolio manager's best interest is in putting some comparative measures in place on portfolio value in order to provide a continued reference point for ongoing decision-making. Most portfolio roadmap timelines extend beyond the quarterly rhythm of earnings reports – frequently well beyond – so portfolio metrics can be useful to align investments to key strategic objectives and help belay the temptation to sacrifice long-term strategy for short-term gain.

Holding the portfolio manager accountable to portfolio value delivery accomplishes multiple objectives:

1. **Portfolio health** – This term can have multiple definitions, but broadly speaking "portfolio health" equates to probability of delivering as per plan. This translates into establishing and tracking metrics that are predictive of the portfolio's overall probability of success (reference Chapter 11: Clinical Development Plan – Section 11.3.5 on PTS and PRS). All the typical project metrics revolving around on-track adherence to schedule, scope, and budget (in dollars and FTE) are certainly all relevant. Additionally, as discussed above, issue and risk rollups are relevant as are relative project weights and inter-dependencies. Individual projects sometimes carry particular importance beyond their budget or explicit benefit, as they have commonalities with downstream projects which magnify their individual weight. Also of importance is the nature of the metrics that the portfolio tracks and publishes out – graphical depictions of predictive metrics such as "buffer tolerance" and "budget actuals versus planned-to-date" are more useful for tracking portfolio health than simple red/yellow/green indicators, for example.
2. **Portfolio execution** – In addition to the other accountabilities listed, most portfolio managers are responsible for the basic mechanics of running the portfolio process. This typically includes a monthly summary status report, current management dashboards, issue/risk follow-ups, and monthly/quarterly meeting reviews. The twin principles which prop up the entire portfolio process are *data clarity* and *focused decision-making* – the portfolio manager thrives or dies according to both. Sections 5.3 and 5.5 below describe in detail each of these key principles supporting the portfolio process.
3. **Portfolio delivery** – Beyond the forecast of delivery for the next several quarters, these metrics are more of a rearward look and seek to measure how well the portfolio has actually delivered against the portfolio roadmap over the last few quarters. Is the portfolio making deliveries according to plan? Is the portfolio keeping its commitments to the enterprise regarding

scope, budget, and timing? Are the various constraints around dollars or resources causing undesired delays?

4. **Portfolio performance** – Viewing the portfolio delivery from another direction, performance metrics inspect how well the portfolio's individual components (i.e., projects) are delivering against their planned benefits (e.g., quarterly revenue, annual cost savings, incremental customer satisfaction). When compared to the benefits promised in the past, how well are the portfolio assets delivering now? As an example, let's say that three years ago, the portfolio team approved development of three new assets – Alpha, Delta, and Epsilon. Each asset would presumably have a specified justification for its development – likely in terms of addressable patient population, ramp up of doses produced, and forecast sales revenue. For this discussion, suppose each of those projects commercialized in different quarters last year. The portfolio manager's job is to compare the actual performance for each of those products against what was projected. Do the actual metrics fall short of projection? A likely culprit could be due to delayed milestones, issues in development, or possibly overestimated benefits.

   Extending the example a bit in order to compare concepts – if the projects delivering Alpha, Delta, and Epsilon delivered their promised scope, on budget and schedule, then the portfolio delivery metrics would all show positive, but if any product failed to meet planned revenue or patient quality of care targets (for example), then the portfolio performance metrics would show negative.

5. **Portfolio contribution to strategy** – These metrics should go beyond planned benefits and should track how the portfolio is contributing to achieving overall corporate strategy. Given the 3- to 5-year objectives for the enterprise, how well is the portfolio supporting those objectives? Is the company evolving its technology platform? Or is the company improving its method of therapeutic delivery? Is the company perhaps looking to expand into new markets (reference Chapter 7: Managing International Projects) or indications? The projects that are accountable for attaining these objectives need to be tracked for progress against these objectives, and the portfolio manager should be held accountable for that progress (or lack thereof).

5.3 WHAT DOES THE PORTFOLIO MANAGEMENT PROCESS ACCOMPLISH?

As highlighted above, if a product portfolio is expected to be truly strategic, then the portfolio manager should hold at least partial accountability for achieving the long-term objectives of the enterprise. The challenge, of course, lies in the fact that long-term enterprise objectives usually take years to fulfill – so what is the portfolio process expected to deliver meanwhile in the current year, quarter, or month? Portfolio execution can be summarized in two key concepts – *data clarity* and *focused decision-making*. While data clarity is covered well below, we will go over focused decision-making first (Figure 5.3).

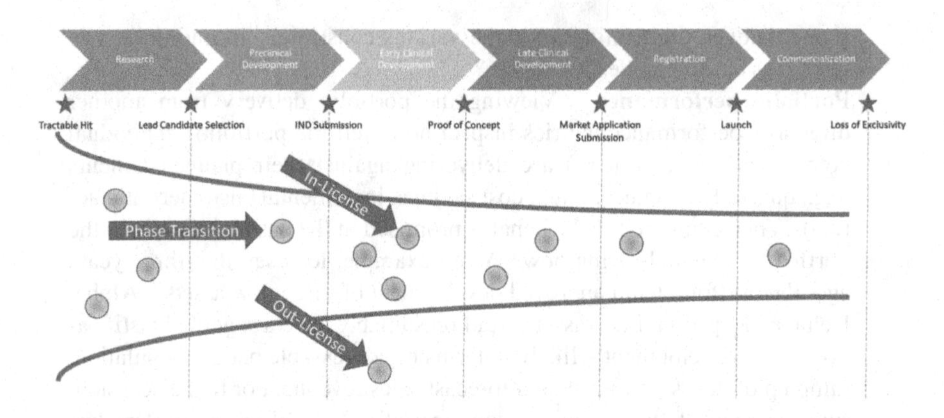

FIGURE 5.3 Representation of the common decision points in Project Portfolio Management.

5.3.1 Portfolio Management and Focused Decision-Making

Regardless of how enterprise success is defined – revenue growth, customer satisfaction, value delivered to shareholders, employee engagement, impact on society and stakeholders, etc. – the net impact of that success is tied to decisions made by the people within the enterprise. How a company expands, who is hired, how the company is financed, how a product or service functions, who it serves, where it is deployed, how much is invested are all decisions made by people looking to grow the enterprise, and all the decisions driving product/service offerings are at the core of driving enterprise success. While some form of governance typically covers any decision made across the leadership landscape, the process which addresses the creation and deployment of products and services of the enterprise is the typical domain of the product portfolio. Thus, while the product portfolio process is a convenient location for sharing metrics on status and issues, the core function of the portfolio process must remain trained on focused decision-making.

Innumerable decisions comprise the development of a product or service, and the ultimate decision to invest the company resources to bring a product or service to market typically rests with the executive committee (reference Chapter 18: Decisions). This team represents the core functions of the enterprise and helps bring the cross-functional perspective in determining the right places to invest for future growth. Numerous tradeoffs are involved in selecting the right direction for new products and services – patients with unmet need, payers/providers involved, future strategy, revenue projections, cost estimates, margin forecasts, complexity/risk, etc. – and only a team with a truly cross-functional perspective can properly assess all the risks versus opportunities across the enterprise, to best judge which are the most advantageous investments to commit at any given time.

Since the portfolio management process is accountable for driving cross-functional decision-making, the process must provide clarity on the roles of the participants, the problems which need solving, and the information needed to support that problem solving activity. The portfolio manager, as facilitator of the process, should be constantly driving to ensure that summarized current project/program information is published on a regular basis, and review meetings are held on the agreed upon frequency. The agenda for each review meeting must prioritize the timely decisions which need addressing, as well as the risks, impacts, and urgency surrounding each. As part of the agenda building, the portfolio manager should also be clear on the type of decision that is being made – is it purely a yes/no decision to move forward with an opportunity, or is it an either/or decision to resolve a tradeoff? In many cases, individuals or teams may wish to address the portfolio team in order to test ideas, seek guidance, provide updates, raise concerns, etc. If left unchecked, this approach will result in the review team expending inordinate amounts of time with little impact on the course of the portfolio. Thus, one of the portfolio manager's most important jobs is in curating the topics that the executive portfolio team will be working on, and ensuring that the time spent in meetings is used for effective, focused decision-making.

5.3.1.1 Portfolio Management Process Areas

The project portfolio management (PPM) process itself has numerous interpretations across myriad practice professionals, tools, and consultants. Approaches vary in format and level of detail, depending on approach and perspective. Some PPM process models define the process as a typical process flow, in a step-by-step fashion, but the more useful models define the PPM process more as a framework of iterative subprocesses that each run continuously and interdependently. Whereas a project is an entity with a defined beginning and destination, the project portfolio lives on in perpetuity. The goal of the PPM process is to simply continue increasing the portfolio predictability and value and hopefully enhance the stated objectives as highlighted in the previous section.

For demonstration purposes, we will use the CREOPM framework[1] created by Dr. Richard Bayney and Ram Chakravarti, as an example of an iterative structure that can be applied to any portfolio of projects. Please note that the discussion here is not intended to be an exhaustive exploration of this framework – please consult their formative book (referenced below) for a more thorough understanding. The subprocesses of the CREOPM framework include the following:

Categorize – It encompasses all areas of research and exploration, including new applications of existing technical platforms, as well as discovery of entirely new platforms. New opportunities are usually identified and triaged based on conversations between researchers and product marketing people. Within this step, the process confirms what is Must Do versus May Do versus Won't Do.

Analyze risk – Since any future outcome carries an amount of uncertainty, every project must be evaluated according to controllable and uncontrollable risks. Performing this essential analysis helps evaluate the relative magnitude and timing of expected benefits and provides the necessary calibration for later steps;

Evaluate – It examines each opportunity according to standardized criteria. Typical criteria include benefits, risk, cost, and time and can also include

[1] Enterprise Project Portfolio Management: Building Competencies for R&D and IT Investment Success by Bayney and Chakravarti.

additional assessments covering strategic, competitive, or operational factors. Meticulous completion of this step is important for several reasons. First and foremost is laying a reliable foundation for the follow-on steps, but a more important reason is maintaining participants' support of the process. By ensuring that assessment of the projects is equitable, rigorous, and unbiased, participants' faith in the results (and outcomes) can be maintained.

Optimize – This step frequently iterates so closely with the prioritize step; the two can be thought of as a joint activity. Using the metrics captured in Evaluate, stack the current and new opportunities into various scenarios, balancing for cost and risk, to maximize overall portfolio benefits.

Prioritize – The prioritization process is actually an ongoing operation, with the prioritized project stack typically formally reviewed with executives every quarter or every-other quarter. Perform tradeoffs of weighting of criteria between strategy, risk, revenue, and other benefits, then highlight those opportunities which are worthy of potential acceleration toward commercialization.

See Section 5.4 for a discussion on differences in approach between early- and late-stage portfolios. For late-stage projects, the four key pieces of information that are needed to evaluate a project in the context of the rest of the pipeline are the following:

1. Scope – as defined by the target product profile (TPP)
2. Launch date – as defined by the project schedule
3. Cost of development – as defined by the project budget
4. Risk – as defined by the PTS (reference Chapter 16- Risks)
5. Sales forecast

Manage – Regularly reassess all project criteria based on continued updates from the project status reporting process. Produce ongoing forecast updates on project external spend, scope, resource needs, and schedule, and make adjustments as needed.

As part of the preparation for the review, for each project in the portfolio (existing and proposed), the portfolio team needs to

- Update total forecast cost (breakout external cost and resource cost/FTE)
- Sum total actual costs (breakout external cost and resource cost/FTE)
- Sum revenue forecasts
- Summarize expected market and delivery risks
- Compare time-to-market *versus* forecast cost *versus* forecast revenue *versus* risk profile for each project (note that various charts can all be useful for this analysis – column, X vs Y, and bubble). Expect to use whatever charts are necessary in order to reveal helpful insights during your executive review sessions
- Create a proposed prioritized list of projects based on the analysis above
- List the prioritized projects, including their time-to-market, forecast cost, forecast revenue, and risk profiles, also total cumulative cost (external spend and resource) for each incremental project on the list. Also be sure to represent any cross-project dependencies (e.g., prerequisites)

- Using the cumulative cost and resource columns, indicate which lower-priority projects are "below the line" and not fundable without additional investment. Conduct the review session with the executive decision team and adjust the list according to executive input, staying mindful of all constraints and dependencies. Continue iteratively until a list of projects is achieved which represents the balanced executive team perspective.

5.3.2 Decision-Making Approach

Over the course of human development, humans have developed a love/hate relationship with making decisions. From one perspective, decision-making is a part of everyday life and is an integral part of any endeavor. Making decisions is how we make progress against our intended goals and thereby have an impact on our world. Progress and impact (especially visible impact) are frequently how we measure success for ourselves and others at work and in life. For these same reasons, however, decision-making also carries a certain amount of risk – since no one can be sure regarding the outcome of our decisions, we are typically concerned (sometimes impactfully so) with the potential downsides of our decisions. Why are we so worried about the potential downsides? For the very reasons why we are incentivized into making decisions in the first place – progress and impact. A poor decision can lead to reduced progress and negative impact, which affects our reputation, credibility, and future prospects (this works for both career and life).

So – in order to make ourselves feel better about possible negative outcomes, we frequently look for additional inputs (e.g., data, opinion, and research) in order to allay our fears about an upcoming decision. This serves multiple outcomes. First, it allows us to avoid making the decision now, by telling (or deluding) ourselves that we will be in a more informed position to make the decision later. Second, it also enables us to hope for reduced negative outcomes by telling ourselves that the overall risk will be reduced with this additional input. Third, it equips us with additional psychological protection (from ourselves and others both) so that we won't be all alone should the decision turn out badly, because we tell ourselves that the people behind the additional inputs will share in the blame too.

What does all of this have to do with project portfolio management? Lots. Recalling to mind the primary purpose for the PPM process is about focused decision-making, and then, the portfolio manager has an ultimate duty to continually drive the important decisions which impact the portfolio. The challenge for the portfolio manager in this pursuit is to continually mitigate the inertia that delays or dilutes effective decision-making. In some cases, projects or programs may need executive guidance to move forward, or a prioritization decision between programs must be made to direct resources that are in contention, or a tradeoff must be made regarding scope that will have an impact on strategy and/or future revenue. In all cases like these, if the choices were obvious, the decision would make itself – what the executive team is required to do is make the decisions which are not obvious, which carries inherent risk of being wrong, and takes process, data, and time. The greatest benefit the portfolio manager can provide to the decision-making is to work closely with the decision-makers and defuse any possible objections raised to making the decision now versus later.

5.4 HOW DIFFERENT PORTFOLIOS ENGAGE WITH PROJECT TEAMS

Any business working to bring innovations to market has the challenge of evaluating investments at different levels of maturity, and pharmaceutical companies carry some of the biggest challenges in this regard. While many of the techniques and examples provided in this chapter are most easily applied to later-stage development projects, the portfolio process can be applied to projects at any stage so long as the criteria are clear and relevant to the stage in question. Research and early development (R&ED) projects are different from late development (LD) projects for several reasons, and they should be treated differently. For these reasons, companies with this dilemma frequently apply different governance to the different groups of projects – effectively treating R&ED and LD as separate portfolios.

The value use of PPM techniques in evaluating the R&ED pipeline is challenging for several reasons. First, the TPP and project plan for an R&ED project is full of uncertainties because the registration path has yet to be defined. Second, the traditional measurements used to perform portfolio analysis (e.g., DCF, NPV/eNPV, IRR, and ROI) rely on sales forecasts that, for an R&ED project, are wild guesses at best. Third, the probability of success for early projects is so low that risk adjustment usually results in a negative eNPV. Therefore, other criteria are typically used to evaluate the R&ED pipeline which includes science-based parameters such as the predictability of the animal model or validation of the molecule target.

While evaluating the targeted benefits for early-stage projects can be a challenge, the other areas of portfolio analysis are still highly relevant. Tracking portfolio spend versus budget, prioritization, resource allocations and tradeoffs, risk analysis, and milestone management are all methods which can help maintain visibility to portfolio health and viability of the expected benefits. Arguably, maintaining a vibrant executive discussion around the state and direction of the R&ED portfolio is one of the best ways to ensure alignment with the company strategy.

5.5 TYPICAL PORTFOLIO AND PROJECT DATA EXCHANGE

As discussed above, an effective portfolio process must be trained squarely on effective decision-making – decision-making which ideally increases future portfolio value. In order for the decision-makers to feel capable and secure in making decisions, they need not just data, but real insights regarding the current and future state of the portfolio to help guide their decisions.

5.5.1 Current State Metrics

Expected portfolio outcomes are aligned to the future, but in order for executives to envision the path to that future, they must have a clear understanding of the present. Any navigation from "Point A" to "Point B" needs a clear understanding of the location of "A" as well as "B". Thus, the role of the portfolio manager is to curate the right number of metrics that help describe the overall current-state health and viability of the portfolio, as well as the future state. This requires establishing standardized metrics and qualitative reporting regarding the portfolio's component parts – that is, the projects.

What is portfolio health versus portfolio viability? Portfolio health seeks to depict the current state of each of the underlying projects, whereas portfolio viability seeks to depict the likelihood of those projects to deliver against the stated objectives. Typical metrics for tracking current portfolio health include rollups for the standard project status reporting – overall red/yellow/green status, current scope percentage complete, schedule timeliness, schedule estimate to completion (ETC), and budget consumed. Typical metrics for reporting portfolio viability include rollups of project issues, risks, forecast schedule changes, forecast scope changes, forecast budget, and forecast resource needs.

Be aware that true project status and outlook (and thus the overall portfolio view) may be masked by limiting the review to only standard metrics; thus, leveraging more detailed metrics may prove more insightful. For example, many projects use "Percent of total resource hours consumed" as a proxy for "Percent complete". This approach can be misleading if a project currently has several major tasks underway and does not have true clarity as to when the tasks will likely deliver (despite what the current schedule says). By using techniques such as asking task owners how much resource time is required to deliver the rest of the task (i.e., not just ask for an expected delivery date), your underlying project status reports may provide better insight as to their future outcome.

Another technique that can prove useful for assessing project schedule is looking closely at the number and size of project tasks that are past due, not just if the major milestone dates are changed from the latest baseline. Task owners are rarely provided formal instruction on how to estimate task schedules. Sometimes task owners can assess the level of completion for current tasks more easily than trying to predict the exact dates for future milestone completion. Recognizing that a task's future completion date has slipped sometimes amounts to admitting to failure for task owners, which can then encourage them to maintain irrational optimism about recovery of the task schedule. Such optimism, if culturally commonplace, can affect significant portions of a project portfolio and delay delivery of committed objectives, that is, clinical trial milestones, quarterly revenue targets, quarterly cost targets, and annual growth targets.

A third approach that can help assess true portfolio health involves assessing external actual spend and seeks to identify not just what the project has spent, but if that spend has accomplished all that was expected. In some cases, services or materials that are procured by a project do not meet expected standards, and the project will need to incur a variance in order to fully accomplish what was scoped. By ensuring that the project manager has fully integrated the issue management and risk management processes with the budget forecast process, the portfolio manager can ensure that the budget ETC figures rolled up into the portfolio budget forecast are indeed a reflection of reality, rather than simply a math calculation of "Total Project Budget" minus "Project Spend to Date".

5.5.2 Future State Metrics

The stated objectives (and criteria) of the project portfolio should form the basis for the forward-looking metrics presented during any portfolio executive review. For example, if projects are selected for portfolio inclusion based on how they contribute to future growth of patients treated, then the portfolio review should include modeling for total patients treated over the next 3–5 years. If projects are selected for portfolio inclusion based on contribution to future revenue or net profit, then those metrics should be included in the model as well. In some cases, the objective may be more strategic – contribute to enhanced treatment delivery methods, or expand existing treatment platforms, for example – and therefore more difficult to measure. In this case, simply representing the delivery of the new capability on the portfolio roadmap can be useful for showing the executive team what the future state will look like.

Put simply, the use of future state metrics is meant to provide the executive review team with a depiction of the "Point B" – in other words, a framing illustration of where the portfolio is headed. Note that utilizing scenario modeling techniques can help executives get a better feel for what the future state looks like and allows them to apply some sensitivity analysis to see which types of tradeoffs (project accelerations, decelerations, inclusions, or exclusions) will have the greatest impacts on desired scenario outcomes.

5.6 RESOURCE DEMAND AND CAPACITY MANAGEMENT

The portfolio management process is responsible for ensuring that the projects across the portfolio are ultimately achievable and deliver value, which usually requires frequent assessments as to whether the projects have the resources needed to deliver effectively. Projects frequently fluctuate in their resource requirements due to schedule delays, scope changes, or planning errors, and many times, new projects need additional resources that were not originally planned – these are all examples of how a portfolio can experience fluctuations in total resource *demand*. Employee attrition, transfers, and role changes are all examples of how a portfolio can experience fluctuations in total resource *capacity* (please see Appendix 5 for definitions of resource management).

Resource management at the portfolio level, therefore, seeks to balance total resource demand with capacity, looking for potential gaps before they become an issue to project delivery. As an example, suppose the project manager for Project Alpha is anticipating a need for a research assistant in 3 months, and this demand is not currently in the forecast. Once the project manager adds the new demand to her project schedule, she would then turn to her core team for (a) confirmation on the size and timing of the demand and (b) assistance in locating resource capacity to address this demand (see Chapter 4). In some cases, the resource constraint is not apparent to the individual project planners – for example, after noticing that multiple projects are all forecasting a major regulatory filing in the same month, only the portfolio manager would be in the position to highlight the risk of regulatory resources being at 98% of capacity in that month. If the necessary resource capacity is not readily available, then the issue would be raised to the portfolio level to either (a) adjust competing project schedules to accommodate the demand, (b) adjust competing project scope to accommodate the demand, or (c) add resources, if time permits.

Facilitating these types of decisions truly underscores the art and science of the portfolio manager's role. Driving the necessary tradeoffs between projects competing for resources is one of the primary jobs of the portfolio manager. Doing this frequently requires insightful charts of analytics and information regarding current status, demand forecasts, risk, and expected benefits for all competing projects. Unless the portfolio team proactively chooses the path for the portfolio, the resource constraints will inevitably cause the competing project(s) to be delayed in reactive fashion, which will eventually demoralize the teams as they start feeling unable to succeed against an unrealistic project schedule. Projects typically are all better served by keeping a realistic plan that all members retain faith and commitment in supporting (for more on this, see the section on pros and cons of annual goals in Chapter 15).

5.7 THE ROLE OF FINANCE IN THE PROJECT WORLD

For a smoothly operating portfolio and financial plan, the relationship between Portfolio Management and Finance must be close and collaborative. Portfolio Management provides clarity of roadmap and associated risks to the financial plan, and Finance provides data and rigor back to the portfolio. Touchpoints between these two processes typically include confirming project costs, benefits, major milestones, and resources (see Typical Finance and Project Data Exchange below).

5.7.1 The Portfolio Roadmap and Associated Risks

The ultimate output of the portfolio management process is a time-based depiction of the project deliverables, or a portfolio roadmap. This graphical schedule layout enables the entire enterprise to align as to the timing of future product releases and helps each department – Research, Development, Nonclinical, Clinical, Regulatory, and Production – ensures that it has its respective resources prepared for when the project needs them. The portfolio roadmap similarly enables the Finance team to align its multi-year revenue, budget, and resource assumptions with the planned product deployments and associated risks.

The project team has a critical role in supporting the finance team by providing project schedule, budget, resource, and risk estimates to the portfolio management process that are accurate and current. The portfolio manager rolls up all the project estimates into a summarized view that enables the finance team to understand what scope is delivered when, the forecast benefits, the cost to get there, and the uncertainties involved.

5.7.2 The Long-Range Plan and Fiscal Budget

The long-range plan (LRP) is typically a 3- to 5-year plan (sometimes longer) reflecting all the high-level activities expected to happen over that period. Finance develops extended headcount and cost growth assumptions based on that plan and also lays out extended revenue targets on that plan. This planning function is essential for ongoing executive strategic decision-making, attaining revenue targets, and overall corporate governance. The project manager role engages with the LRP process by identifying major anticipated milestones within the LRP time horizon, which then enables Finance to create a long-range budget and headcount forecast.

The annual fiscal budget, sometimes called the annual operating plan (AOP), while related to the LRP, is different and is the deliverable that most projects engage with. As it is aptly named, the annual fiscal budget is limited in scope to the fiscal year and is generally much more detailed (and prescriptive) toward headcounts and dollar budgets. Resource and budget owners that do not stay within their defined limits are generally restricted from doing so or may be subject to disciplinary action. Projects, which must have resource and dollar budgets in order to thrive, are strictly limited by the headcount and budget restrictions that the annual budget places across the company. As part of the budget allocation process, all dollars and headcount are usually earmarked for specific projects or other purposes. Thus, projects which run into overruns and thereby require additional dollars or FTE frequently need to have them reallocated from other projects or activities. For these reasons (and others), project managers are asked to accurately estimate their tasks, schedules, FTE, budgets, and risk, in order to ensure that the overall totals stay within the annual business plan. When a project requires more time, FTE, or money, the portfolio management process is called upon to re-balance the portfolio according to the executive prioritization.

5.7.3 How Finance Engages with Project Teams

Finance is frequently called upon to engage with a project team:

1. At the beginning of the project to confirm its assumptions on benefits, spend, and risk; and
2. Throughout the project to track the budget versus actual spend (aka BvA, or budget variance analysis).

The nature of the interaction between Finance and the project team can vary with company culture, but the intent is largely the same. While the project team is singularly focused on product creation and deployment, Finance is focused on ensuring that the company maintains its fiscal obligations, so the project can continue its work year after year. Prior to launch, the project team creates numerous estimates on FTE cost, external spend, benefits, and risk which many times get reviewed by Finance to ensure validity of assumptions and consistency across projects. During project execution, Finance is usually involved in reviewing the actual project dollar spend and compiling the revised spend forecast for each project. This governance helps ensure that projects are held accountable for their estimates and that the company manages its money according to its intended priorities.

Some individuals may view product development and Finance as operating at cross-purposes – one team looking to expand revenue and the other seeking to constrain it. In reality, both are seeking the same overall goal (prudently growing the company), just from different perspectives. Product development seeks to bring innovations to market as quickly as possible, and Finance seeks to ensure that the two components of expected ROI (i.e., the investment, as well as the return), are both kept within acceptable ranges. By maintaining a healthy essential tension between these two teams, judicious product investment and resulting company growth can be accomplished.

5.7.4 Typical Finance and Project Data Exchange

As highlighted above, the conversation between Finance and Project Management is normally conducted within the domain of data and analytics. Taking the forecasts, assumptions, and actuals produced by a project, Finance can help assess project risk and how well the project fits within the broader annual and long-range plans.

The data that typically flow between the project team and Finance includes the following:

Milestone delivery schedule – Revenue forecasts produced by Finance are largely based on commercial delivery assumptions contained within the project schedule. For this reason, accuracy in project planning is essential. Poor assumptions which lead to project delays not only result in greater cost for the project (thereby impacting ROI) but also adversely impact revenue outcomes for the target fiscal year;

Resource FTE estimates – The project schedule (including resource task assignments) can provide a short- to intermediate-term forecast (6- to 12-month window) for project resource needs. The project team typically also provides additional long-range resource forecasts (12- to 36-month window) that are based on high-level activities, such as initiation of new programs, data readouts, and regulatory submissions. Finance can typically use these long-range estimates to then derive their own overall resource cost forecasts, supporting the annual budget and long-range plan;

External spend estimates – In addition to resource estimates, projects also communicate to Finance their external spend estimates. The costs typically associated with external project spend involve materials and equipment supporting internal research and services supporting project delivery. Maintaining a baseline version of the budget and comparing that to later revisions can help both parties track the justification for budget changes over time;

Actual external spend – The accounts payable system tracks external spend and tabulates that spend based on project, category, and timing. Financial reports back to the project assist the project manager in tracking spend to ensure that the project is keeping within its allocated budget. By combining actual spend with remaining spend (i.e., estimate to complete or ETC), the project manager can accurately forecast quarterly budget expectations;

Commercial forecasts – As the project moves into clinical and commercial phases, more in-depth revenue forecasts are produced to help justify the project resources and spend. Once reviewed and vetted by senior leadership, these forecasts form the foundation on which Finance bases the enterprise planning and budgeting process.

Risk assumptions – Since project outcomes and deliveries are uncertain, the project team carries a crucial responsibility to document its assumptions around risk and thoughtfully consider what the impacts of those risks could be on scope, budget, and schedule. Will the expected therapy provide the expected patient results? Will the amount of resources or spend required possibly increase? Will the expected therapy possibly be delivered late? Close partnership between the project and Finance on the risk analysis is crucial to ensure that both teams can accurately depict the uncertainty surrounding the costs, timing, and impact of planned deliveries.

5.8 SUMMARY

The key concepts supporting the portfolio process can be summarized in this way:

- The project portfolio, simply being a rollup of projects and programs, carries the same data objects, constraints, and aspects as projects do. The portfolio seeks to balance scope, budget, schedule, and quality the same way that projects and programs do. The portfolio carries risks and issues the same way that projects do. The portfolio also delivers promised benefits the same way that projects do. One major difference is that since the portfolio is a **rollup** of projects, it is similarly a rollup of all the project aspects that are contained within it – scope, budget, schedule, risk, issues, and benefits. The portfolio, however, is more than the sum of its parts, as it is ultimately accountable for helping the enterprise deliver against its strategy. This translates into taking the above project aspects, integrating the overall risk tolerance of the company, and applying prudent investment decisions that achieve the company's needs (e.g., revenue, cash flow, market objectives, customer growth, and targeted strategy) in the both short and long term.
- The portfolio manager, similar to the project manager, is responsible for maintaining the deliverables under their responsibility, and for tracking scope, budget, etc., but at a higher level. The difference between the two is only the level of purview. For both the project and portfolio processes to run effectively, the portfolio manager should be focused on health and viability at the portfolio level and leave the running of the projects/programs to their own managers. An effective analogy is how the human brain relies on the rest of the subsystems of the body to operate effectively – the brain does not directly control the internal functions of the liver and kidneys, but is nonetheless wholly reliant on their success.
- The data exchanged between projects and the portfolio process are about monitoring and maintaining health and viability. Dashboard rollups regarding delivery status, spend vs budget, resource, risk, etc., help the portfolio committee stay aware of current state and enable them to make better decisions regarding prioritization and future innovation. Thus, status reporting of metrics from projects to the portfolio process is not so much about command and control, but enabling visibility for executive insight.

The triad of Portfolio Management, Project Management, and Finance Management forms a crucial coalition that plans, tracks, and delivers value to the enterprise. Portfolio managers ensure that the right projects are selected and prioritized; finance managers ensure that appropriate funding is available to deliver projects; and project managers, working through the project team, deliver the projects.

Part 2

Contemporary Topics in Drug Development

6 Agile Project Management and Its Application to Drug Development

Kamil Mroz

CONTENTS

6.1 INTRODUCTION

> It's not the strongest of the species that survive, nor the most intelligent, but those most adaptive to change
>
> **Charles Darwin**

DOI: 10.1201/9781003226857-8

Since its inception in 2001, Agile has gained popularity as a project management method due to its success in some of the largest companies in the world (e.g., Uber, Apple, Spotify, Airbnb, Alibaba, and Facebook). Despite its success in the tech sector, some industries, including the biopharma sector, are still struggling to embrace its full potential. This may be because drug development, while being a volatile, uncertain, complex, and ambiguous (VUCA) environment where Agile can thrive, is unique in its scientific and regulatory context. Nonetheless, project managers in biopharma can still benefit from knowing the Agile principles and being able to apply Agile practices to appropriate situations during the drug development process. This chapter aims to provide development project managers with the history, concepts, use cases, and best practices of applying Agile to the drug development process so that we can leverage this methodology as one of the tools in our toolkit.

6.2 WHAT IS AGILE AND WHY SHOULD A PROJECT MANAGER CARE?

There's a good chance someone has asked you in the past 5 years "Are you using Agile?". Truth is, most people don't really know what Agile is. The trendiness of the term has taken on a form of its own, and the singular definition of Agile has been convoluted by myriad applications and branded models. To understand Agile and how it applies to development project managers, it is useful to know the origin of the methodology, the difference between "doing Agile" and "being agile", and the business context in which Agile is most useful.

6.2.1 History of Agile

In the late 1990s, software developers were becoming increasingly frustrated with the low customer satisfaction in their newly released products. The problem was that a new software project would follow a set plan to deliver a predefined set of features and specifications, but during the 2–3 years it took to create the product, new technologies were being introduced that changed the customers' expectations, and so the new product no longer satisfied the customers' expectations. Using traditional project management approaches, development teams did not have a mechanism to capture the customers' changing expectations and build responsive features into their product before release. Thus, the need for a new way of managing projects arose.

In 2001, a handful of software developers met to "uncover better ways of developing software".[1] For more information on the origin of Agile, there's a great article by Caroline Mimbs Nyce in *The Atlantic* that gives the whole history.[2] The resulting document, called the Agile Manifesto, laid out 4 values and 12 principles to shape an iterative and customer-centric approach to software development.

[1] http://agilemanifesto.org/.

[2] https://www.theatlantic.com/technology/archive/2017/12/agile-manifesto-a-history/547715/.

Four values of the Agile Manifesto
• Individuals and interactions <u>over</u> processes and tools
• Working software <u>over</u> comprehensive documentation
• Customer collaboration <u>over</u> contract negotiation
• Responding to change <u>over</u> following a plan
Twelve principles of the Agile Manifesto
1. Customer satisfaction through continuous delivery of the product
2. Divide large chunks of work into smaller and achievable tasks for quicker completion and easier integration of changes
3. Adhere to the decided timeframe for the delivery of a working product
4. All stakeholders must frequently collaborate to ensure that the project is going in the correct direction
5. Create a supportive environment to motivate team members and encourage them to get the job done
6. Prefer face-to-face communication over other methods
7. Working software is the primary measure of progress
8. Try to maintain a constant pace of development
9. Maintain the quality of the product by paying attention to technical details
10. Maintain simplicity
11. Promote self-organization in the team
12. Regularly reflect on your performance for continuous improvement

Over time, software project managers transformed the principles of the Agile Manifesto into project management methodologies (e.g., SCRUM, SaFE, XP) and practices (e.g., minimum viable product – MVP, iterations, Kanban) that became the norm in their sector. An extensive set of concepts and terms arose, as evidenced by the extensive list in Appendix II. The enormous success of tech companies in the early 2000s then inspired other industries, including biopharma, to try to apply Agile to their project management practices.

6.2.2 What Is Agile and How Does It Compare to Waterfall?

The core concept of Agile Project Management is to use an iterative approach to developing a product. Rather than pre-specifying a long list of features and requirements that are needed for a final product and then taking a long time to achieve all those features and requirements, Agile aims to produce an MVP and then repeatedly build upon that MVP with incremental improvements based on feedback from end users. Thus, the product evolves over time, which is why Agile is considered an adaptive approach to project management.

Waterfall, on the other hand, is considered a predictive approach. Waterfall project management begins with a clear definition of the product and a clear plan on how to build it. If the plan is fully executed, you should expect to get a product that meets all the pre-specified requirements with all the expected features. Waterfall follows a sequential development path (Figure 6.1).

While waterfall and Agile appear at opposite ends of the spectrum in terms of flexibility and adaptability, there may be a blend of these two approaches into a hybrid methodology that works well for drug development. Development project managers

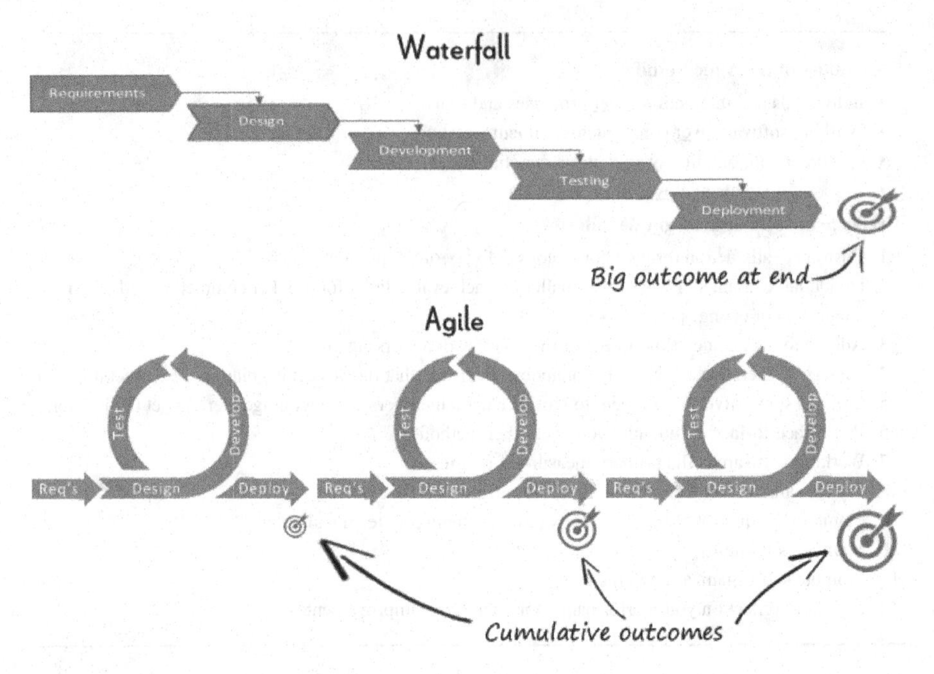

FIGURE 6.1 Representation of Waterfall and Agile project management methodologies, highlighting the early delivery of value and iterative approach to product improvement in the Agile approach.

should have a working knowledge of all these approaches so they can pull the right tool from their toolkit based on the specific project situation.

Predictive approaches are necessary for hypothesis-testing situations in drug development. For example, a pivotal clinical trial is governed by a protocol that sets out to prove or disprove a hypothesis (i.e., an objective). Adaptive approaches may be used in situations where iterations are allowed, for example a dose escalation clinical trial that tests a dose level and gathers information to support a decision to reduce the dose, stay at that dose, or increase the dose.

When should you use Agile vs Waterfall? A project manager should use his/her judgment. If your project is similar to another project that has been executed many times, there is a clear process, and the output is clearly defined; then, a waterfall approach may be more appropriate. However, if your project has many uncertainties, requirements may evolve, and the outputs are not well defined, then using Agile can be beneficial. The Stacey Complexity Model uses two parameters (clarity of requirements and clarity of process) to guide the use of Agile vs Waterfall (called "Traditional" in the diagram) (Figure 6.2):

6.2.3 "Doing Agile" vs "Being Agile"

There are a lot of misconceptions about Agile that are important for development project managers to be aware of. First, it is important to note that "doing Agile" is not the same as "being agile".

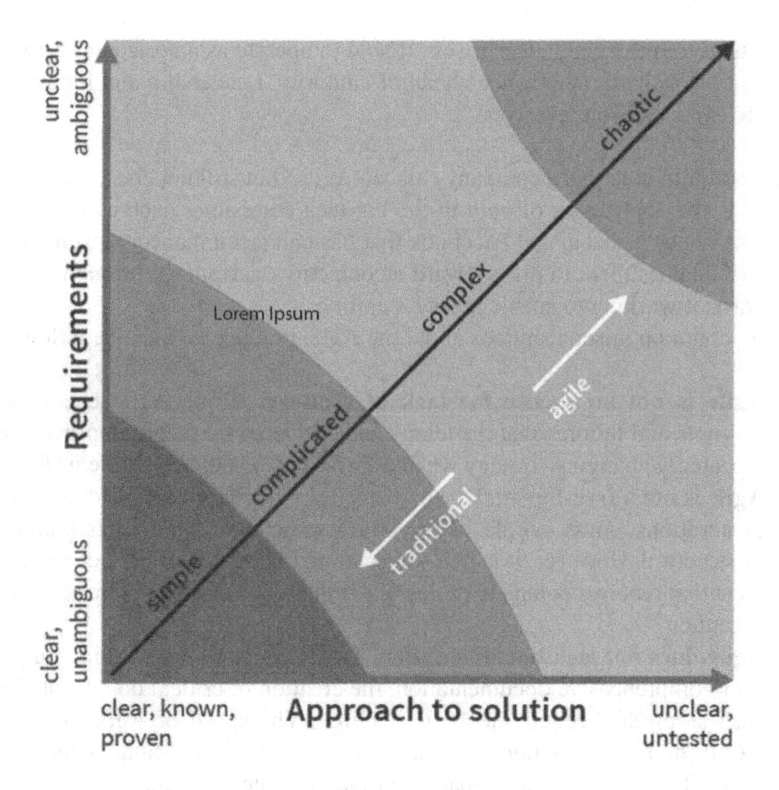

FIGURE 6.2 Schematic illustration of the Stacey Matrix.

- *Doing Agile* means adopting the principles, values, methods, roles, processes, and tools of the Agile Method into one's project management practice. In this book, we'll call this the **Agile Practice**.
- *Being agile* is an organizational mindset of being nimble, flexible, and reactive to the changing external pressures. In this book, we'll call this an **agile mindset**.

Steve Denning, in a 2020 Forbes article[3], nicely summarizes three concepts of an agile mindset:

1. **The ideology**: agile enterprises are all obsessed with delivering value to customers as the be-all and end-all of the firm; profits are the result, not the goal of the firm.
2. **The architecture**: agile enterprises all tend to organize work, to the extent possible, to be carried out in small self-organizing teams, drawing on the full talents of knowledge workers and working short cycles.

[3] https://www.forbes.com/sites/stevedenning/2020/05/24/doing-agile-right-from-agile-mindset-to-agile-principles/?sh=186d89685fab.

3. **The dynamic**: agile enterprises all tend to operate as a horizontal network that prioritizes competence ahead of authority. Leadership and innovation are expected from everyone.

It's important to note that a company can "do Agile" but still not "be agile", and vice versa. It is the application of *both* in the big tech companies such as Uber, Apple, Spotify, Airbnb, Alibaba, and Facebook that has enhanced their organizational success. This enables them to pivot toward a competitive advantage through the use of Agile ways of working to enable business agility.

Other common misconceptions about the Agile practice include the following:

- **Agile is not an excuse for lack of strategy**. While Agile encourages attempts and failures that the team can learn from for future improvement, repeatedly changing strategy when a sprint fails is not how Agile is done.
- **Agile is not a free-for-all**. Due to its focus on flexibility and on doing work in iterations, some people incorrectly assume that Agile lacks planning or structure. However, it is still a disciplined approach to project management that requires complete project planning for each sprint or phase before execution.
- **Agile does not lack documentation**. While Agile favors working outputs over comprehensive documentation, the creation of critical documentation, such as product requirements, is required. The speed of Agile comes in part from the information flow that comes from daily coordination meetings rather than lengthy reports that take time to create and read.
- **Agile is not an excuse for poor quality**. Just because your next iteration is in two weeks doesn't mean the current one is allowed to be insufficient – Agile teams are still expected to provide a workable product at the end of each phase.

6.2.4 VUCA Environments and How Agile Can Help

Agile practices and the agile mindset provide the most benefit to volatile, uncertain, complex, and ambiguous (VUCA) work contexts. The table below defines VUCA.

VUCA Component	Definition
Volatile	The environment often faces unexpected changes or issues that have a major impact on project success.
Uncertain	The environment offers little information to confidently make high-stakes decisions.
Complex	The environment has many dynamic, interrelated parts and variables.
Ambiguous	The environment lacks specificity, with information that can be interpreted in many ways.

VUCA environments require true project management, and Agile practices and the agile mindset can help to minimize disruption in these environments. The following table describes how project managers can apply Agile methods and the agile mindset to minimize disruption in VUCA environments.

VUCA Component	Opportunities for PMs to apply Agile practice and the agile mindset
Volatile	Project managers can address volatility by helping the team to pivot due to unexpected changes or issues. Pivoting can be done through work reprioritization and refinement. Work reprioritization relates to changing the priority of remaining work for a particular work stream to maximize the value of the effort being put into that work stream. Refinement is a regular and dynamic process that helps teams shape well-sized, detailed, and discrete pieces of work that can be tackled in future iterations.
Uncertain	Project managers can address uncertainty by being explicit about the assumptions made during the planning process. As these assumptions are tested through experiment or observation, the assumption is changed to fact and knowledge, and the team gains more certainty.
Complex	Project managers can address complexity by breaking down dynamic, interrelated work into manageable scope. In addition, the Agile practices of collaboration, transparency, continuous learning, knowledge sharing, and "adaptability to change" in the project environment are all intended to reduce complexity.
Ambiguous	Project managers can address ambiguity by helping the team assess the robustness of the information they have and helping to bring more clarity. Additional clarity can be obtained through collective "expert judgment" available in the cross-functional team structure that Agile promotes.

Drug development in general is a VUCA environment. Though many aspects of drug development are regulated and not amenable to Agile practice, there are some areas where Agile can be applied effectively. Furthermore, there are many areas where the agile mindset can be applied. We'll look at some examples in the section below.

6.3 WHERE CAN AGILE BE APPLIED IN THE DRUG DEVELOPMENT PROCESS?

Historically, development project managers have utilized the waterfall project management methodology because the scientific method lends itself well to the stepwise approach of the waterfall method. However, the success of Agile in many industries has led many biopharma companies to evaluate the use of Agile in drug development.

In an HBR article published in 2019, the authors describe their experience with implementing Agile practices in a variety of science-driven businesses, from drug development to chemical manufacturing[4]. In the article, they describe their work to implement "Agile science" – a new way of working characterized by the pragmatic and context-specific use of Agile methods and tools. Applying Agile science to these companies netted benefits like a 20% increase in R&D productivity and a doubling of the capacity to run projects, while also increasing success rates. By integrating

[4] https://hbr.org/2019/11/why-science-driven-companies-should-use-agile?sf113187444=1.

Agile deep within the organizational core, these biopharma companies were able to tackle sudden internal changes as well as the volatility of the external environment.

In this section, we will explore the challenges of translating Agile practices from software development to drug development, an option to combine Agile with other methodologies to create a hybrid approach that works for biopharma and where development project managers can apply this hybrid approach to certain aspects of drug development.

6.3.1 The Challenge of Implementing Agile in Drug Development

Agile practices are not suitable for all activities in drug development, and it's important for development project managers to understand the reasons why before embarking on an Agile implementation. To begin with, it is helpful to consider the differences between drug development and software development contexts. The following table highlights some of the key characteristics of drug development projects compared to software development projects.

Characteristic	Drug Development	Software Development
Project Management Approach	Predefined phases and stage-gates	Flexible iterations and versions
Regulation	Highly regulated	Low level of regulation
Project duration	10–15 years	< 1 year
Project costs	$500M–1B	$100K–1M
Consumers	Patients	Customers

The different contexts may explain why development project managers have struggled to apply Agile practices. The regulated, hypothesis-testing context of drug development imposes inherent barriers to the flexible and iterative Agile project management approach. Let's take a deeper look at some of the characteristics of drug development that directly challenge Agile principles:

- Drug development progresses in a rigidly controlled and sequential manner (i.e., Phase 1, 2, 3). The phases of drug development are set up by regulators to protect patients from being exposed to drugs that do not work or can cause harm. This challenges the Agile principle of gaining customer satisfaction through continuous delivery of the product.
- Biopharma is a highly regulated industry. Quality standards, evidence-based decisions, and patient safety concerns all make it difficult to start with an MVP to release to the consumer and iterate with improvements. Indeed, national health authorities are charged with evaluating the risk–benefit profile of each drug before it is allowed to enter the market, and so project teams try to produce the best possible product the first time. We'll see later in this section that one could consider lifecycle management tactics to be

"Agile-like", albeit at a macrolevel (several years between iterations) compared to software development where iterations can be in terms of monthly iterations.
- The concept of having minimum requirements is disorienting to many scientists and researchers who are trained in the scientific method of first developing a hypothesis and then testing the hypothesis to achieve validation. Although it is advisable in late clinical development to aim for a target product profile (see Chapter 11), projects in early research are really just trying to figure out what the drug is and how it behaves in cellular systems, which is not conducive to setting minimum requirements.
- While software development generally aims to make people's lives more convenient and entertaining, drug development aims to improve human health. This difference engenders a very different risk tolerance in biopharma when compared to software development, with biopharma being much less comfortable with uncertainty and ambiguity. One can imagine the consequences of a glitchy app on your phone being far less impactful than a "glitchy" drug that affects a person's liver or kidney.
- Compared to software projects, drug development projects are an enormous commitment. The long development times and high investment costs make it challenging to iterate and pivot quickly. It also discourages trial-and-error approaches because the trials are so time-consuming and costly.

Context counts, and so one of the major challenges to overcome in bringing Agile to the pharmaceutical sector is trying to force-fit or blindly applying Agile practices without considering the inherent specificities of the pharmaceutical drug development process. One of the major challenges to overcome for pharma is to tailor Agile to its own realities. Therefore, let's try to ask a few pharma-focused questions derived from the Agile values in a way that helps contextualize the right questions to ask when embarking on your Agile transformation:

Individuals and interactions over processes and tools	• How could Agile improve the day-to-day working within an interdisciplinary development team? • What is the composition of your team and how do you focus on working together within your team? • How do you ensure cross-functional collaboration with your team? • Do you have a charter focusing on behaviors, norms, and team values? • When things go wrong do you default to your processes or try to figure things out with your colleagues?
Working outputs over comprehensive documentation	• How do you focus on a "working therapeutic" over comprehensive GxP documentation? • How can you optimize delivery in a highly regulated environment? • How do you deliver to provide increments of values to your internal or external customers? • Is there an MVP that you can release to your customers?

Customer collaboration over contract negotiation	• In pharma, a customer can be a patient, their families, caregivers, communities; payers; or providers – how do you think with them in mind? • How do you obtain inputs from patients, their families, caregivers, and communities to deliver the highest quality of care across settings? • What is an increment of value to deliver to your customers? • How do you focus on the "must-have" requirements or features while progressing a drug through a complex and highly regulated process?
Responding to change over following a plan	• How would traditional development stage-gates evolve to accommodate Agile techniques to respond to change? • How do you focus on developing shorter cycle times to make drug discovery and clinical development more efficient? • How can you use iterative planning to deliver incremental value from research to the launch of a medicine? • How can your team be more adaptive to respond to changing business requirements when developing a new drug?

Despite the differences between software development and drug development, there are still areas within the drug development process where Agile principles can be applied. We'll look at a few of these areas below.

6.3.2 CMC Process Engineering

Agile practices can be applied to the manufacture and process engineering of chemical compounds. When considering the manufacturing plan, a prototype dosage form can be created early using a "crude" process. For example, an initial dosage form may be just the active pharmaceutical ingredient (API) that is hand-filled into a standard gelatin capsule shell. This prototype can be used in initial testing and engineering runs to determine the feasibility of the formulation and dosage form for further development (e.g., adding excipients to extend dissolution, improving compressibility to make tablet forms, and improve flow properties). Thus, the prototype dosage form can be considered an internal MVP.

Furthermore, it is often the case that the initial process used to manufacture a compound can be optimized to produce a higher yield, fewer steps, or more steps with reliable reactions. These process optimization efforts are iterations of the initial process that improve the output, benefiting from feedback obtained from prior manufacturing runs.

Finally, from a quality control perspective, it is often the case that initial manufacturing specifications are loosely set and then tightened up as more information is gained with each run. The initial specifications are purposely set with wide error bars so that batches are not rejected, but as the manufacturing team learns more about the process through in-process testing, it gains more confidence in being able to replicate the process with a tighter range of quality control specifications. While not the same as the Agile principle of continuously evolving the product, this iterative approach can be considered Agile-like in that it continuously provides a similar product while continuing to evolve the way it is produced.

6.3.3 Combination Product and Medical Device Development

When pharmaceuticals and medical devices are combined into a combination product, the project management environment changes. This is especially true when the combination product falls under the medical device regulatory requirement. The regulatory environment for Type I and Type II medical devices (low and medium risk) medical devices is well suited for a hybrid approach that is closer to the Agile style of project management. This is especially true when making relatively minor changes to an already marketed product.

Device development takes a more risk-based approach that does not have much of the hypothesis-testing context of pharmaceuticals. In addition, some simple iterations (such as a minor change in size or shape) only require a simple letter to file instead of a more comprehensive SUPAC update. In addition, some medical devices have software, which is also well suited to the Agile philosophy. More details on project management for combination products are provided in Chapter 8.

6.3.4 Early Development

As described in Chapter 1, the drug development process is broken down into research, early development, late development, and registration. In piloting Agile approaches through this process, I've realized that generally, the research and early development stages are more conducive to Agile approaches because the context allows for rapid iterations with early stakeholder involvement. Moreover, when you break down the research process into its corresponding stages (target ID and validation, hit generation, etc.), you start to get into timeframes that are more conducive to iterative approaches, all of which have their key priorities to get to the next stage. Such a breakdown also allows the project manager to use Kanban and sprints around key deliverables that drive a minimum business increment (MBI).

6.3.5 Lifecycle Management

On a macroscale, lifecycle management tactics can be considered "Agile" in the sense that initial drug products can undergo incremental improvements to become new products on the market. The initial product can be considered the MVP, and subsequent introductions can be considered iterations of that MVP. Examples of this concept include the following:

- The antidepressant Celexa, the racemic mixture of R and S citalopram enantiomers, was improved upon by Lexapro, which contains only the therapeutically active S isomer of citalopram.
- The drug everolimus was initially approved for the prophylaxis of organ rejection after transplantation; then, the label was improved upon with the addition of an indication for the treatment of renal carcinoma.
- The epidermal growth factor receptor (EGFR) inhibitor cetuximab was initially approved in Europe as second-line therapy for colorectal cancer and then improved upon by moving to first-line treatment after clinical trials

revealed that cetuximab plus chemotherapy worked better for patients with the "wild-type" KRAS gene.

- The osteoporosis drug Fosamax was initially approved as a once-daily dose and then improved upon with a new formulation that allowed for once-weekly dosing.
- More recently, one could consider the first wave of COVID-19 vaccines to be the MVP. At the time of writing this book, many more "next-generation" COVID-19 vaccines are in development. These newer vaccines seek to improve the deficiencies of the first vaccines, including broader epitope coverage, longer duration of protection, more convenient storage and handling conditions, and better efficacy against mild infection. Future vaccines may be more suitable for certain age groups, subpopulations (e.g., those with underlying immune-compromising or other medical conditions), and pregnant women.

Although these lifecycle management (LCM) tactics take several years to implement (versus the months it may take for a software application to iterate), they can still be considered as an Agile approach to asset development when looking at the entire brand life. The initial launch can be considered the MVP, and each chemical change, indication expansion, targeted subpopulation, or dosing change can be considered an incremental improvement on that MVP. The MVP allows the drug to reach patients and the brand to generate revenue early in the lifecycle, while the LCM improvement typically enhances the addressable population for increased or extended revenue generation.

6.3.6 Regulatory Submission Development

Agile practices can be applied to the creation of major regulatory submissions like investigational new drug (INDs) and new drug applications (NDAs). The case study below describes a process that uses a SCRUM-like approach to creating content in an NDA. More information about creating a market application submission is presented in Chapter 13.

Case Study: Using a SCRUM-Like Approach to Manage an NDA Development Project

An NDA is often viewed as the culmination of all the work done under a clinical development plan. There can be hundreds of nonclinical, clinical, and CMC reports and datasets that need to be organized, summarized, and integrated into a submission. The process often takes 4–6 months and involves dozens of people from many functions and levels of the organization. A SCRUM-like approach is very useful in creating the submission dossier because of its focus on iterative progression, transparency, prioritization of activities, speed of execution, and customer (i.e., FDA) focus. This case study describes my experience using a SCRUM-like approach to manage an NDA development project.

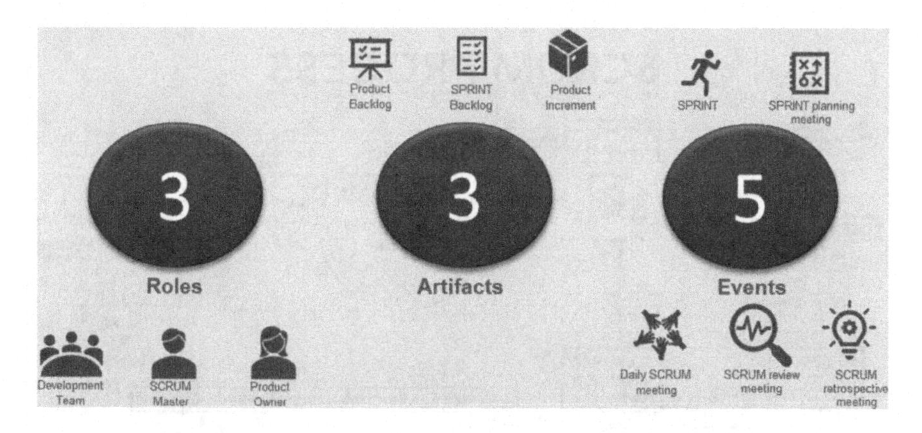

FIGURE 6.3 Typical roles, artifacts, and events for SCRUM.

First, let's discuss some basics of SCRUM. SCRUM is a simple Agile methodology (the guide itself is only 19 pages long), and so it is a good starting point for a team to pilot Agile approaches. The terms "SCRUM" and "Agile" are often used interchangeably, but they are different: *Agile* refers to a set of methods and practices based on the values and principles expressed in the Agile Manifesto, whereas SCRUM is a methodology used to implement Agile.

SCRUM involves three roles, three artifacts, and five events (Figure 6.3).

For the NDA project, I did not want to introduce new terms to the team, so I adapted the terminology to fit what we already understood. You'll see this is a common theme throughout this case. So, the following 3–3–5 definitions were used:

Three roles	Submission owner, project manager, and author/reviewer
Three artifacts	Content tracker, content creation plan, and NDA development plan
Five events	Authoring step, reviewing step, QC step, weekly standup meeting, and messaging alignment meeting

After adopting SCRUM definitions to our business, I then created a SCRUM process that we could use to draft the modules of the NDA. The following figure presents a common SCRUM process (Figure 6.4).

And here is the modified SCRUM process that we used to create the NDA (Figure 6.5):

Throughout the process, MS Planner was used as a Kanban to visualize and track the backlog (i.e., not yet started), ongoing (i.e., in progress), and completed tasks. The Kanban format was chosen because it shows the status of work items in a visual and intuitive way, allowing team members to see the state of every piece of work at any time. MS Planner was available for everyone on the team to use, providing transparency into the assignments and flow of work.

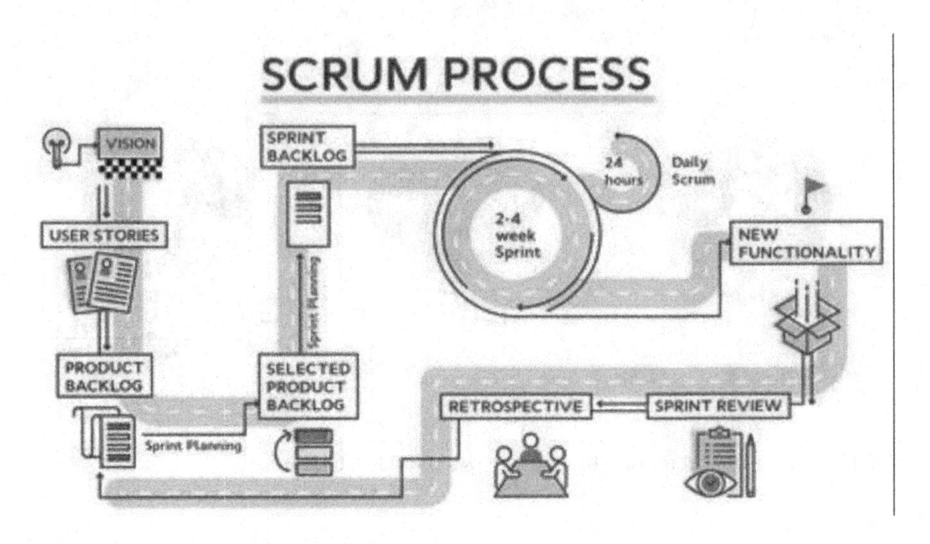

FIGURE 6.4 Generic SCRUM process.

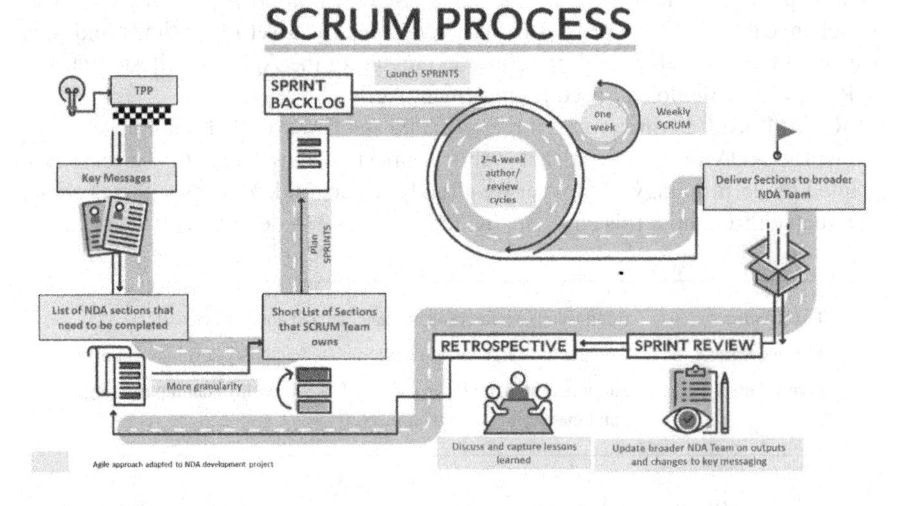

FIGURE 6.5 Modified SCRUM process to manage the preparation of a regulatory submission.

6.3.7 A How-To Guide on Managing SCRUM-Like Meetings

The above case study describes the use of a SCRUM-like process to manage the development of a marketing application submission, and key to that process is running effective SCRUM-like meetings. SCRUM meetings are used to launch sprints and to share progress during sprints, ensuring that the team can collaborate and plan as needed for a productive sprint.

Agile Meetings - Goals and Benefits

Scrum Meetings AKA Ceremonies

Sprint Planning

The Meeting	Tips and Benefits
Purpose: To define a realistic Sprint goal and backlog containing all items that could be fully implemented until the end of the Sprint by the Scrum team. The sprint planning meeting results in two Scrum Artifacts, the Sprint goal and Sprint backlog.	**Who should atte[illegible]** [illegible] Product Owner, Scrum Master and the [illegible] Team.
How it is conducted: • The Product Owner defines the Sprint Goal - a short description of what the sprint [illegible] attempt to achieve, clarifies the details on backlog items and their respective acceptance criteria. • These entries are updated and broken into smaller stories by the team so [illegible] they can be completed within one Sprint. • The stories are estimated, prioritized and tasked. • The Scrum Team defines their capacity for the upco[illegible] - the total capacity of the Scrum Team might change from sprint to sprint [illegible] to provide realistic commitments, it is necessary to know the total [illegible] of the team for the upcoming Sprint, with consideration of vacations, pu[illegible], etc. • Tasks are assigned amongst the team [illegible] based on their experience levels and expertise. • The team is ready to start with [illegible] sprints.	**U[illegible]ful T[illegible]:** • Timebox the meeting. Stop when you reach time. • The Product Owner should have the Product Backlog prioritized and ready before the meeting. Only review stories that are 'ready,' i.e. meets the Definition of Ready (DoR). • Review and agree on the acceptance criteria that says when a given work is considered 'done,' i.e. meets the Definition of Done (DoD). • The Scrum Team selects how much work they can do in the coming sprint based on their capacity and should not be influenced by the Product Owner. • The Product Owner should clarify with stakeholders on any unclear requirements for the team. • Plan for collaboration of team members and NOT for optimal 'resource utilization.'
Benefits: Below [illegible] the benefits of running a successful Sprint Planning meeting: • Enables the [illegible] to agree on the sprint goal and commitment. • Creates the platform to communicate dependencies and identify team capacity to set and commit to an achievable sprint goal. • Ensures that daily sprints remain productive.	

Good example on how to structure

FIGURE 6.6 Example of an agenda for a SCRUM meeting.

6.3.7.1 Step 0: Create a Backlog

A backlog contains a prioritized list of all the key deliverables that are needed to be completed to get to the next step of the process. It can be collected in a Kanban or a spreadsheet. It's important to run this in a cross-functional manner to ensure that key risks, interdependencies, and sound prioritization are done as a team (Figure 6.6)

6.3.7.2 Step 1: Prioritize the Work

Use your backlog to prioritize the work and break it into "manageable chunks", aka sprints. Sprints should be SMART – with a clear goal and time-box. A project manager can help to facilitate the process and ensure the team focuses on the content

6.3.7.3 Step 2: Plan a Sprint Planning Session

Sprint planning sets up the entire team for success throughout the sprint. Coming into the meeting, the product owner will have a prioritized backlog. The objective of this meeting is to kick off the sprint and clearly define the sprint goal. The recommended time is 45 min, so it's important to be efficient during the meeting. The key outputs of this meeting are:

- A clear sprint goal with a time-box (typically 2, 4, or 6 weeks)
- A SCRUM team of key contributors to the sprint, including the project manager, contributing team, product owner
- A checklist of key activities needed to achieve the sprint goal with clear responsibilities captured in the checklist within the sprint card

The project kickoff meeting is to get everyone on the same page and off to a great start. It is an opportunity to introduce the team and increase understanding of the sprint so that work can begin as soon as possible.

6.3.7.4 Step 3: Launch a Recurring SCRUM Meeting

The objective of a recurring (daily or weekly) SCRUM is to coordinate, follow up on, and track the key activities in a recurrent manner using the program team Kanban. The SCRUMs are held until the sprint time-box is completed. Note that during the SCRUM, the team does not go into technical details; instead, it's an opportunity to update the team on progress, plans, and risks. The project panager can use the following three questions to gather inputs from each of the contributing team members:

- What have you done since the last weekly SCRUM?
- What will you do before the next SCRUM?
- Is there any impediment?

The recurring SCRUM is a 15–30-min event for the Agile team to synchronize activities and create a plan until the next SCRUM. The project panager and product owner should be there to help the team focus on risks and barriers and ways to circumvent obstacles.

6.3.7.5 Step 4: After Completion of the Sprint

After completing a sprint, the contributing team will share the output of the sprint with the broader team and key stakeholders in a sprint review. Sprint reviews can be done at the end of the sprint or any time a key output is available that may impact the broader team. In addition, the SCRUM team will hold a retrospective meeting where they gather to discuss and capture lessons learned and integrate them into subsequent sprints. You can very easily just ask the three questions and capture them in the minutes of the meeting:

- What worked well during the sprint?
- Areas of improvement?
- Key take-away actions?

The benefits of this SCRUM-like approach to managing meetings are the following:

- Strengthened accountability among Agile team members
- Agile team is also accountable to improve its way of working
- Agile team is accountable to meet the sprint goal: "They are self-organizing"
- Improved line of sight for project teams on priority focus areas

- Improved visibility into activities that need to be completed via the prioritized backlog
- Greater transparency on the key deliverable of the program
- Sprint reviews create better transparency between Agile team and stakeholders
- Kanban boards allow team members to see task progress at any time
- Participation of project management team as SCRUM masters and owners of tracking tools

6.4 SUMMARY

Project management has evolved since its inception and will continue to evolve. We are living in times of extraordinary change, where society and organizations are constantly being confronted with change and are pressured to stay innovative. VUCA makes it that competition keeps getting fiercer, resources are becoming scarcer, and customers are more demanding than ever. Agile helps us to navigate this volatile, uncertain, complex, and ambiguous landscape. It has really impacted the way project management is being carried out through new knowledge areas, ideas, and techniques that are likely to have a lasting influence. If things continue like this, Agile will become a pivotal reference point in pharmaceutical project management, helping project managers make better decisions, effectively plan, and easily adapt to change.

While agility has recently attracted considerable attention in studies of systems development and management of information technology (IT), there is limited recognition within the pharmaceutical sector. Agile approaches are a new form of organizing that is conducive to innovation. Agile Project Management methodologies will further enable pharmaceutical organizations to be agile in the so-called VUCA environments. Literature is still relatively nascent. Hence, it is important to detect the antecedents of project management methodologies enabling pharmaceutical organizations to compete. Even more so in the case of pharmaceutical companies, as the COVID-19 pandemic has introduced unprecedented challenges and opportunities to this sector where agility can be further studied and applied for the benefit of practitioners and researchers.

In the meanwhile, I hope that through the case studies and practical examples, you have become more aware of Agile and an agile mindset to be able to apply it to your pharmaceutical context – enjoy the journey, and good luck!

7 Managing International Projects

Henri Criseo
Santen

CONTENTS

7.1 INTRODUCTION

International expansion is often viewed as a rite of passage for a successful biopharma company, but one that must come with careful deliberation and often with some trepidation. Expansion is expensive and risky. It requires building new organizational

DOI: 10.1201/9781003226857-9

capabilities, hiring more people, establishing new ways of working, and adapting to unpredictable changes to company culture. Nonetheless, the decision to expand is typically supported by tangible and intangible upsides that align with the company's vision.

This chapter describes some common reasons for international expansion, the challenges that global project teams (GPTs) face, and some recommended practices to make international projects successful.

7.2 WHY DO COMPANIES GO INTERNATIONAL?

Put simply, a biopharma company is willing to invest time and money to expand into other regions because it wants to access more patients. These patients could be part of a development program (e.g., additional patients to enroll in a clinical trial) or part of a life cycle management strategy (e.g., to maximize the commercial value of an asset through additional revenue streams). In addition to these tangible benefits, there is also an intangible benefit in that the company expands its corporate image, thereby stimulating investment opportunities from a broader set of investors. We will go into more detail on these reasons for global expansion below.

7.2.1 Tangible Benefits: Faster Development Programs and Additional Revenue Streams

A common driver for international expansion of a development program is to access more patients for clinical trials. In the United States, international clinical trials really started to be utilized in the mid-1970s when language was introduced to the Code of Federal Regulations (21CFR312.120). Initially, only large companies took advantage of their international footprint to conduct multiregional clinical trials (MRCTs), but nowadays companies of all sizes are utilizing MRCTs either through their own footprint or through global or regional contract research organizations (CROs). A company gains several benefits by expanding its development plan to include international clinical trials.

First, the higher number of countries means a larger pool of patients that can enroll in the clinical trial, thereby accelerating recruitment rates and shortening the study duration. Most sponsors these days will use a CRO that has foreign presence (either via a global or regional CRO), thereby avoiding the need to hire local expertise to run clinical trials. In this situation, although the sponsor study team may be co-located, the project team still becomes an international team, and the project must be managed differently than a localized team. In addition, in the case of a small biopharma company, it is often felt that a global CRO is unlikely to provide their best talent, and so there are advantages to using a regional CRO in terms of customer service.

Second, to receive market approval, regulatory agencies will often require clinical safety and efficacy data from a patient population that is representative of the population it serves and consistent with the medical practice of that country. This is especially true where certain ethnic factors may be unique to a country or region. For example, differences in genetic and dietary differences between Japanese, Chinese, and Western patients have led the regulatory agencies of those countries and regions to require a certain percentage (from my experience, between 10% and 50% depending on the project type, disease area, etc.) of clinical trial participants from those countries or regions. Therefore, if you plan to commercialize your product in these countries or regions, you will need to conduct clinical trials in those countries or regions.

A quick aside here that I'm seeing more often: companies with an existing international footprint often look for ways to take advantage of their presence in multiple global locations. One benefit is that work can "follow the sun", meaning a team member in one time zone can work on a task during his/her workday and then pass it over to another team member based in another time zone when that member starts his/her workday. The biopharma industry has mostly employed this practice for manufacturing, data management, and statistical programming activities, and it is starting to be used in other areas such as Medical Writing and Clinical Operations. Another benefit is that resources can be shifted to regions with lower labor costs. For example, Clinical Operations teams can reside in China, Poland, or Canada where labor costs could be lower than in the United States and the United Kingdom.

From a commercial perspective, startup companies will seek international markets to maximize the value of their development efforts. Indeed, in order for a company to maintain growth, it needs to look outside of its country or region of origin because local growth could plateau after just a few years on the market. For US-based companies, expansion into Europe and Asia is the common next move. For companies based outside the United States, a move into the US market is usually the first move due to its immense market size. Thus, global expansion of the commercial capability typically arises around the time of the first launch of a company's first product. Although rare, some companies will seek an immediate global launch. More commonly, international expansion occurs in waves, with the most valuable markets being in the first wave.

Let's consider the following case study. A device maker with international presence successfully launched a product in the Europe, Middle East, and Africa (EMEA) region and then looked to expand into China, where the patient population is almost twice the size of that of the EMEA. The project team felt the probability of success was high for China because the EMEA region was already happy with the product (good feedback from patients, key opinion leaders acknowledging the product's quality and efficacy, no major safety concerns, etc.), and the overall expected net present value (eNPV) and return on investment supported the move. So the project team decided to propose the project to its affiliate in China. It took several months to think through the product positioning in China and its integration within the company portfolio in the country, and the clinical study design that would fit with local regulations, but eventually the company decided to move forward with development in China. This staggered approach enabled the company to gain revenue in the EMEA region first before pursuing other markets. That being done, this would become a virtuous circle as the company will continue developing its knowledge locally in terms of product development and strategy as it would have paved the way with new products, starting interactions with local competent authorities, and so on.

7.2.2 Intangible Benefits: Broader Corporate Profile and Increased Institutional Knowledge

From a perception standpoint, international expansion is seen as the management team's confidence in the product they're developing and a commitment to company growth. By increasing the visibility of its core business, a company can often take advantage of broader opportunities for fundraising and seeking strategic business partners.

Another benefit of having an international team is the gain of institutional knowledge of local regions. Having "boots on the ground" is often viewed as the best way for the folks in corporate headquarters to have the best understanding of foreign local markets and regulations. For example, developing a product in Europe is challenging because of many country-specific practices and regulations. It is often beneficial to have local experts who can help navigate the clinical trial startup and execution activities in each country.

7.3 HOW DO INTERNATIONAL PROJECTS AFFECT A DEVELOPMENT PROJECT MANAGER?

For project managers working at large companies, the international footprint is typically already in place and the global team structure is usually already running. International projects will usually have a global project manager (GPM) and one or more regional project managers (RPMs). For example, in a US-based international company, a GPM may sit in the US headquarters and liaise with RPMs in Asia-Pacific, Europe, or South America. The GPM is accountable for integrating regional development plans into a global development plan. Thus, project teams are dispersed by time, geography, and culture, which poses logistical, lingual, and cultural challenges for the project manager to create an environment where project work can be done effectively.

7.3.1 Logistical Challenges

International project teams face logistical challenges due to physical separation, time zone differences, and differing working calendars. We will discuss each of these aspects in detail below.

7.3.1.1 Physical Separation

Having a team that is physically separated by countries or oceans means that members are not able to meet face to face as often as co-localized teams. This became even more true after the COVID-19 pandemic, when international travel ceased altogether. The obvious implication is that "chance encounters" in the hallway or kitchen area are nonexistent. Therefore, the GPM needs to have other ways to maintain contact with foreign team members.

Fortunately, digital communication tools greatly improved in the late 2010s, making correspondence more casual and quicker across international barriers. As the COVID-19 pandemic spread across the globe in 2019, many companies increased their reliance on videoconferencing tools such as Microsoft Teams and Zoom. However, these tools have their limitations.

While sound or video quality is not perfect by far, the major limitation of digital communication tools is both intangible and critical: trust is built through stronger relationships, and in-person relationships are stronger than virtual ones. Teams that physically sit with each other during meetings, share an after-work dinner, or just bump into each other in the hallway have an enhanced level of trust and a sense of psychological safety, which are the hallmarks of a highly effective team.

In most cultures, sharing a meal with someone is the first step in building a relationship that goes beyond the formal corporate tone, to talk about everything else than business and share experiences and thoughts, and this is difficult to apply when doing virtual meetings only. It's also in these kinds of exchanges that you discover the local cultures and get to know how people interact and how to make it the right way by listening, looking at a specific environment. These casual face-to-face interactions have the effect of turning the voice on a call or the face on a videoconference into a colleague or even a friend. This has simply disappeared while doing video conferences only: time is spent most often working on projects, plans, issue resolution, risks assessment, etc. When joining a new international team with 15 people from different cultures, no one will guide you through the local traditions, so you'll have to figure it out another way.

Even if these interactions have been considerably lowered, one thing we should keep in mind as a project manager is to build the foundations of a team the right way, and this has to go through knowing the people who are part of the team. Often, the project manager is most qualified to do so since each project manager normally has great communication skills to foster team interaction.

I recommend organizing 1:1 video calls with team members. This helps not only to get to know your colleagues a bit more, but also to get news from each other that may affect the project, outside of team meetings: for example, what personal situation may be affecting your colleague's ability to deliver, any issue he or she might face, any help you could provide. Doing so shows that you really care, and this enhances the overall trust between each other. It's not necessary to spend hours doing 1:1 video meetings, but it is good to schedule them at a regular cadence. And they don't need to be formal with an agenda and objectives; typically just talking about business and projects will naturally lead to sharing of new information.

As a matter of personal preference, I prefer to activate the video as much as possible. Seeing facial expressions, though not as complete a picture as full body language, is helpful in being able to understand nonverbal communication. You won't get as much nonverbal feedback as if it was a real face-to-face meeting; however, it is still better than making voice-only calls. Nonverbal feedback is really key to getting 100% of the message shared with you by your colleague, and you can better comprehend what is the overall meaning and sense of a verbal conversation. Listening and interpreting your peer's silences are also an important part of active listening as we are going to discuss further.

7.3.1.2 Time Zone Differences

Time zone differences could have a big impact on the GPM's ability to organize meetings or events within the project team globally. For example, consider a part of a global team is working from the West Coast United States and the other part is in

China, which means there will be 9 hours of difference; thus, while offices in one part will start their business day, those in the other part will be close to finishing its working time, and this has to be taken into account in the communication, team dynamics, and expectation from others.

Such time differences are limiting the window where both offices could be connected, and thus, this needs to be anticipated. A GPM should organize team meetings accordingly taking this into account.

For example, if he or she is based in the western part of Europe (Paris, Geneva, etc.), the morning time shall be devoted to communicating with Asia as a priority (e.g., China, Japan, Southeast Asia); then during the beginning of the afternoon, calls could be set up with the US East Coast and finally the US West Coast by the end of the afternoon, all the while not forgetting to communicate with his or her local region the rest of the time, but this would normally be easier during the rest of the day. One should keep this kind of agility and flexibility while communicating to ensure having enough room and time to connect with other regions at the right time.

Such an exercise becomes even more complex when three or more regions across the world with big time differences need to be connected at the same time. While having the great honor to manage such kind of "super global" projects as a GPM, it can (and will) become painful from time to time knowing that at least one of the regions will have to take calls in either early morning or late evening.

To overcome this potential issue, the whole team should first be made aware of it from the very beginning of the project by the GPM setting the scene by explaining who is sitting where in the world. Then, the GPM should organize the core team meetings in such a way that each region should not be impacted every time a meeting occurs. This means that meeting hours should rotate, allowing each team member to participate during his local business hours on a regular basis.

From my experience, not doing such a rotation will exclude some of your colleagues who will be obliged to constantly join during unconventional hours, which can cause disengagement in the long term until they simply become unresponsive to your requests. This practice may also be perceived as a lack of respect in some regions, as colleagues could eventually feel not being listened to or appreciated. In the long term, this can impact employee retention.

To simplify and support such meetings in an organization, we can suggest several tools to find the most suitable meeting time such as a meeting planner (available at www.timeanddate.com/worldclock/meeting.html), which allows you to pick several dates and cities globally to set up calls; you can also add extra time zones within Outlook Calendar, add several clocks through the iPhone widget, or wear a GMT watch, etc., so that anytime you need to schedule a meeting you simply know what time and day it would be for your counterpart (Figure 7.1).

7.3.1.3 Holidays and Working Days

Another thing to take into consideration is to respect each country's local culture about national holidays and working days. If you work with Japan, there are 16 national holidays and festivals during the year which the Japanese celebrate; thus, it's important to know when these occur.

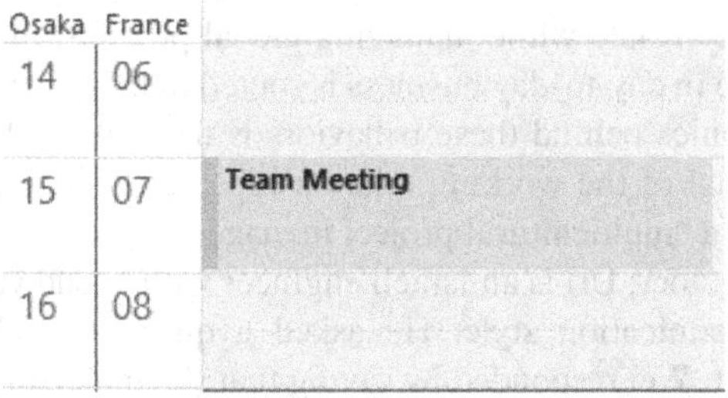

Osaka	France	
14	06	
15	07	Team Meeting
16	08	

FIGURE 7.1 Example of Outlook calendar where time zones in France and Japan are displayed.

Although the Monday-through-Friday work week is fairly consistent across the world, there are some regions that adopt a different schedule. For instance, several countries in the Middle East work Sunday through Thursday in order to observe Friday as a Muslim holy day.

7.3.2 Language Barriers

Although most companies have adopted English as the common language for business interactions, the level of fluency will vary from individual to individual on your team. A language barrier has the obvious implication of ineffective communication, but it also has the subtle implication of preventing a sense of community on the project team. Establishing a common language is key to sharing information and to getting the team to work together.

People who are more comfortable with the language are naturally more willing to speak and share information. It is essential for a project manager of a multinational project team to recognize the "speaking time" of the relevant participants for a given topic. To include all the right inputs for a decision, the project manager may need to solicit inputs from a team member who feels too shy to speak up.

A project manager will often need to serve as a "translator" for members of the team who are less fluent in the common language. One way to do this is to paraphrase and repeat back what you understood from a member's contribution to the conversation. Another way is to actually learn a few words in your peer's language to show your true desire to communicate. Learning standard greetings, regardless of your pronunciation, will show you want to make some effort to get closer to your project members. Recently, one of my Chinese colleagues welcomed me during a team video meeting by speaking French, I was really surprised since I did not know she could manage to say a few words, and I was really pleased she wanted to raise our level of connection that way (especially since my Chinese pronunciation is just horrible).

7.3.3 Cultural Differences

While sometimes it seems the physical separation of a staircase can be as significant a barrier as an international flight, international project teams have unique challenges in the form of cultural differences between team members. Knowing how to manage cultural diversity is key while managing global projects: the way colleagues or partners will behave in day-to-day business is rooted in their own culture, and understanding the mechanics behind these behaviors is mandatory if you want to maximize all the potential of the working relationships. Let's elaborate further on why you should become a "multicultural project manager".

Let's start with a story. Uri is an Israeli engineer with a loud voice and a "straight-to-the-point" communication style. He asked a question of Wei, a soft-spoken Chinese counterpart. Wei responded by saying that she understood the question. Uri pressed on, demanding an answer right then and there. Wei remained silent. One could hear Uri becoming infuriated. The standoff was immediate and palpable, and it was all due to cultural differences.

7.3.3.1 Working Styles

It's difficult to define, but we all know a different working style when we see one. And while never true of every individual in a category, stereotypical work types have been identified that can be helpful to understand how team members from different regions might think about each other. For example, the pace of business seems to be different across cultures. The Japanese and South Americans sometimes seem to move too slowly for the Americans, and the pace of American businesses seems unnecessarily hasty to the Japanese and South Americans. It is important to know what to expect from team members in other regions (e.g., where a 3-day turnaround time for a complex email is normal), and what they might think about your work style as you interact with them (e.g., sending a follow-up email after 1 day can be perceived as being "pushy").

Also, consider this example. In some cultures, it is acceptable to jump right into the business with new colleagues, whereas in Chinese culture, it is important to first build trust through a personal relationship. The concept of "guanxi" underpins the importance of relationships and social networks in order to facilitate business interactions in China. Once you establish a guanxi with someone, perhaps from having been able to help him or her on a specific matter, it will bring you some advantages such that one day he or she will return the favor to you in another context where you will need help.

Differing working styles can affect how individuals on a global team work together. Passive vs. aggressive, fast vs. slow, and conservative vs. risky – all these perceptions can cause friction on a project team if not carefully managed.

7.3.3.2 Communication Styles

It is important for a project manager of an international project team to understand the difference between low-context and high-context cultures. Low-context cultures rely more on the exact words being used to communicate a concept, whereas high-context cultures rely more on the situation around the discussion than on the actual words being used. Examples of low-context cultures are Americans, Germans, and Israelis. In contrast, high-context cultures exist in Japan and China.

To do so, it is key to give the floor to all participants, in particular to those who did not express themselves yet: consider why they did not talk and ask them to give their opinion on a specific topic while also gently refraining from the "too talkative" members from monopolizing the discussion. Once that person has expressed his/her position, then you can rephrase or summarize what has been said and ask something like "is my understanding correct?". Doing so will (a) confirm common understanding on the topic, (b) show your interlocutors you care about what is said and about their opinion, (c) confirm you want to engage all members within the project team, and (d) make you a trustable project manager as you balance the discussion.

From a pure day-to-day interaction perspective, you will also remark that communication and expression flow are different: the Japanese culture is mainly driven by collectivism, by the importance of respecting "group harmony". There is a will in Japan not to enter too much into conflict nor to challenge each other too much, contrary to a Western straightforward approach where someone will easily express his/her disagreement. This means your Japanese colleagues might sometimes keep their feelings inside (which they call "honne" 本音) rather than express them directly ("tatemae" 建前), in particular to avoid raising conflict and keep harmony within the group.

It is often seen that my best colleagues, even after knowing me for a long time, will use a very polite way to tell me I might be wrong and that they would like to get more arguments from me. Instead of sending me a pure "no", they will prefer using a "maybe" or a "let me think about it" or also "it's difficult" (which in translation essentially means no).

7.3.3.3 Decision-Making Styles

The way groups make decisions is heavily influenced by the local culture. The concept of power distance is a cultural dimension that describes how people behave when the boss is in the room. In cultures with low power distance, subordinates feel more empowered to make decisions; in cultures with high power distance, subordinates usually defer decisions to their manager. American culture is generally considered to be of low power distance, whereas Japanese culture is considered to be of high power distance. I have worked with many Japanese teams during my career, and, though this is evolving in Japan with the arrival of CEOs having a multicultural background and who have spent several years in other regions building other local expertise, the phenomenon still exists.

Decisions made in a Japanese organization typically respect the hierarchy and seniority of managers and leaders. Decisions are not made until the senior-level person agrees to them, and you definitely do not want to bypass the right senior-level person. Also, the organization is sometimes more vertical compared to what might exist in the United States or Europe. Instead of going right to a department head, it is often better to explain a specific issue to a lower-level team member and allow him/her to escalate it up the chain.

In Japan, the concept of "nemawashi" describes the decision-making process where a consensus is reached after long hours, if not weeks, of discussion. Compared to the Western culture, this often takes longer than you might be used to, and you must remember to be patient. You have to engage with all key stakeholders, sometimes one by one, to bring your ideas with a lot of granularity and to explain why you would like to proceed in such a way. Those stakeholders will then share their opinion so you will be able to address their concerns, allowing them to ultimately agree with your proposal to move forward or to invest in something specific. Of course, to do so, you will have to take the time to identify those key people in each function, not omitting any especially outside R&D (e.g., Finance to get agreement on a specific budget, Human Resources to ensure you'll get the right resources, Portfolio Planning to validate a new project value is adequate and fitting with the pipeline). Doing so, you will gain alignment with your ideas and will get support from them. This means that the day you present a proposal to a committee, often the decision would have already been taken.

7.3.4 Differences in Development Pathways and Regulatory Requirements

For many project managers in pharma/biopharma, the initial project experience focuses on a single country region, especially our own home country. We become very familiar and knowledgeable on local practices and requirements. For example, we become familiar with the relationship with the competent authorities, such as for scientific advisory bodies and ethical committees, local requirements for shelf life for investigational medicinal products (IMPs), the market application submission and review process, and so on. However, what is acceptable and common in one country might be different in another country. Therefore, rather than apply the knowledge and expectations we learned in our home country or region to other countries or regions, it is wise to relearn the local practices and requirements without prior bias. Let's take a couple of examples to illustrate this.

First, let's consider the acceptability of study design or endpoints by the different agencies all around the world from a regulatory standpoint. Indeed, the prerequisites to get a product approved by the US FDA might be different from what the EU's EMA or Japan's PMDA require. For instance, if we look at glaucoma, the US FDA will usually require the tested product to be non-inferior in intra-ocular pressure versus the comparator at three different time points during the day, whereas the EMA will accept the peak and trough measures only of intra-ocular pressure during the day. Obviously, these different requirements will affect the way the study is designed in terms of the schedule of assessments and the way the data are analyzed and reported.

A second example could be dry eye disease. Until now, the US FDA approved only drugs that demonstrate superiority versus vehicle on both signs (e.g., corneal fluorescein staining improvement) and symptoms (e.g., eye dryness) of the disease in separate pivotal trials, whereas Japan's PMDA has approved dry eye products based only on a demonstration of superiority versus vehicle based on signs alone.

The two examples above show the differences that could occur from one region's authority to another, which makes it challenging to create a global strategy for several regions at the same time. One region may consider a study to be confirmatory, but another may deem it to be only supportive if it does not meet the prerequisite expectations on endpoint or study duration.

Also, the required duration of pivotal studies for approval could change from one region to another. This is also something that you could take into account while creating your schedule or ultimately evaluating the time to market and the eNPV of your project. A company could consider performing MRCTs to accelerate time to market, taking into account all the key differences across regions needed for a drug to be approved.

Similarly, the regulations for labeling IMP require different information printed from one region to another, making it challenging to orchestrate when the project team would like to minimize the IMP manufacturing costs and clinical labeling by having only one set of text for an MRCT. Also, IMP stability required by competent authorities to start a trial might differ from one country to another; therefore, checking the shelf life "rules" with your Chemistry, Manufacturing, and Controls (CMC) lead locally is mandatory prior to initiating your trial.

As a final piece of advice, I encourage you, the project manager, to partner closely with your regional Regulatory, Clinical, and CMC team members to confirm the local requirements to initiate a trial and ultimately what would be needed with regard to primary efficacy endpoints, study duration, and finally to ensure you build the right development plan integrating all the assumptions locally.

7.4 RECOMMENDED PRACTICES FOR MANAGING INTERNATIONAL PROJECTS

Despite the logistical, lingual, and cultural challenges of managing an international project, there are both organizational benefits and personal rewards that make international projects exciting to work on. The recommendations below will hopefully diminish the difficulties to the point where the rewards can be fully experienced.

7.4.1 Form Geography-Based Team Structures

For global projects, there may be a GPT and one or more regional project teams (RPTs) from, for example, Japan, China, North America, or Europe. The GPT comprises the regular functional and subteam leads plus the RPT leads. Each RPT may have additional subteams (e.g., Clinical, CMC, Regulatory) created at the discretion of the RPM and RPL to support the implementation of the regional development plans.

7.4.1.1 Global Subteam Leads

From a team structure perspective, as a project manager, you must keep track of and disseminate key information to the whole team in a timely fashion to ensure the project is progressing as effectively as possible. To do so, you'll have to rely on your subteam leads who will have been identified when creating the project team roster. These individuals will be the members of the global core team, accountable to the core team and to the GPM for identifying and delivering specific items related to their own function: for instance, you will identify a CMC subteam lead, a regulatory subteam lead, and so forth. Those subteam leads should have a global overview of their activities and functions and will become the reference and key contact point for you. By doing so, you ensure that you keep communication at a very high level and install an effective communication channel for sharing information.

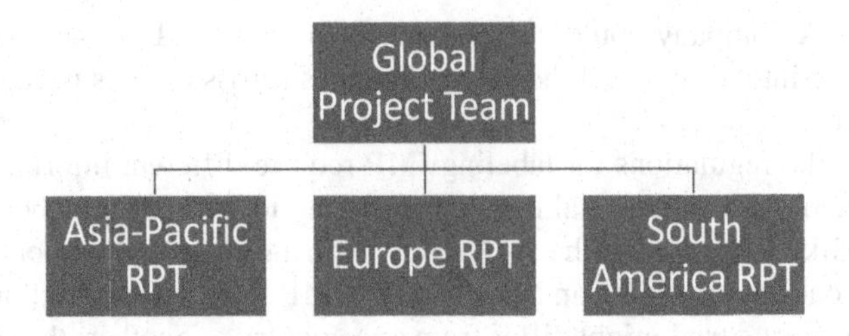

FIGURE 7.2 Example of a global project team structure with regional project teams (RPTs) reporting to a global project team.

7.4.1.2 Regional Project Teams

Sometimes when projects have several regions and when project complexity is increased across regions, you will need to establish RPTs under the core team. Indeed, it can become too complex to manage regional-specific aspects at the core team level only, or the core team might lack the required local expertise. Therefore, building a more local team that will report and be accountable to the core team will make sense. This might be the case if you manage a program in which you have to set up several trials in different regions (e.g., EU, USA, China, etc.), and as explained previously, you'll still have your subteam leads doing the oversight per function (Figure 7.2).

7.4.2 Be Culturally Sensitive

I sometimes hear *I don't like to work with people from this particular country, they don't understand the way we work.* To this, I always have the same response: Did you try to step back and analyze the way you communicated previously? Did you identify the blocking points? Isn't it because you have different habits in the way you work or different expectations?

The first thing a project manager will need to do is to forget the common idea that the working style they're used to in their culture is the only way to get work done and accept there will be differences between the way you work and how others may work in other regions. Then, you will need to adapt your working style and be flexible with how to move your project forward. When you accept this, you can start increasing your level of cultural sensitivity with respect to working styles.

Development project managers, as the mediators of project team dynamics, need to be especially attuned to cultural differences. I recommend learning about local habits in day-to-day working, traditions, and environment when dealing with other regions and countries. You could do so while interacting with your peers in the project team, asking them about their working habits, share your working habits and pieces of your culture to increase your respective levels of curiosity. This will not only help to learn about working together but will also show a real mutual interest and empathy about the way they have been working. This will be perceived by your colleague as a sign of interest, ultimately increasing the level of trust in the group and mutual understanding.

There are many international business courses available that teach about cross-cultural business relationships, awareness of cultural differences, and how cultural differences in other countries influence business and social life. Interesting yourself in others' working styles will dramatically increase your understanding about behaviors you might see in the project team and will help to: (a) accept those behaviors; (b) adapt your behaviors and responses while dealing with cultural differences; (c) help others to also understand how you work and manage projects, and; (d) ultimately improve your business relationship and project team dynamics.

7.4.3 Enhance Communications

As with any project team, communication is vital. It is even more important with international teams because of the language and cultural differences described previously. The following recommended practices will enable the project team to keep information flowing and benefit from the broader pools of experience and expertise.

7.4.3.1 Organized Information Sharing

An international project with a team spread across the globe means that people will be working at different times of the day. To keep information flowing across this type of team, the project manager should set up a convenient space for asynchronous discussion and real-time access to documents and information. This can be done by creating a shared cloud location such as SharePoint or Box. Whichever solution is used, the project manager should be thoughtful regarding the folder structure and organization of information so that it makes sense to the rest of the team when they are trying to locate files. The project manager may want to take time to explain how the file-share environment works and the rules to use it efficiently. Let's be tolerant with project members who have limited experience with sophisticated IT tools; they will appreciate your help to ensure they can spread their ideas to the rest of the team.

7.4.3.2 Regular Interactions

It can be exhausting for a GPM to keep lines of communication open for international project teams, and sometimes a GPM needs to spend long hours with subteam leads across regions early or late in the day. But this is what's required to keep information flowing as real time as possible and to resolve issues efficiently. As the GPM, I take responsibility for sharing information from one regional team to another, especially since they may not interact with each other often due to challenging time zone differences. In addition, international travel to meet other team members is part of the GPM role. This can sometimes be challenging to balance work and personal life, especially for GPMs with families.

7.4.3.3 Active Listening

During global team calls, actively listen to your colleagues, carefully following what he or she means, and how their contribution should be incorporated into the purpose of the meeting. GPT meetings are valuable not only because they're hard to schedule, but also because they have a wealth of insight to draw from. To maximize the opportunity, here are a few things to set up to ensure an environment of active listening.

- Mute notifications such as new emails and chats, and silence your phone
- Have some paper handy to jot down key points, actions, and decisions
- Raise open-ended questions to dig further into topics that need more detail
- Show interest and appreciation for your team members' comments and suggestions
- Confirm your understanding of the discussion by reiterating or summarizing the key points, actions, and decisions from each topic.

Active listening will enable your meeting to conclude with the right takeaways and action plan. Take the full benefit of being 100% connected with your colleague during this time, and observe the nonverbal signs he or she will show to validate the exchange is going the right way and that you are both in agreement. This will lead to very effective communication.

7.5 SUMMARY

Despite the challenges of international expansion, many companies will still pursue opportunities to operate in other countries and regions. To meet the needs of international projects, GPTs will be formed, and with that, new ways of working will need to be adopted. Successful global project teams find a way to address the time, language, and cultural differences to optimize the delivery of their project. Recommended practices include forming a geography-based project team structure, utilizing the local expertise, and enhancing communication techniques as well as using the latest communication tools to be more efficient and increase communication accuracy. When done effectively, an international project has the potential to deliver a new treatment to a broader set of patients who need it.

8 Managing Combination Drug–Device Projects

Rhonda Peck
Cook Biotech Inc.

CONTENTS

8.1 INTRODUCTION

The FDA defines a combination product as "a product composed of any combination of a drug and a device; a biological product and a device; a drug and a biological product; or a drug, device, and a biological product". While device development has parallels to drug development, there are enough unique aspects of the regulatory environment and the required development components to warrant a separate chapter describing their development. In addition, there may be aspects of device development that a project manager (PM) can learn from to apply to drug development.

As the FDA's definition suggests, combination products can come in many forms. Combination products can be single-entity products such as a pre-filled syringe, co-packaged products such as a first-aid kit that includes bandages and pain relievers, or individually packaged products that cross-reference each other for safe and effective

DOI: 10.1201/9781003226857-10

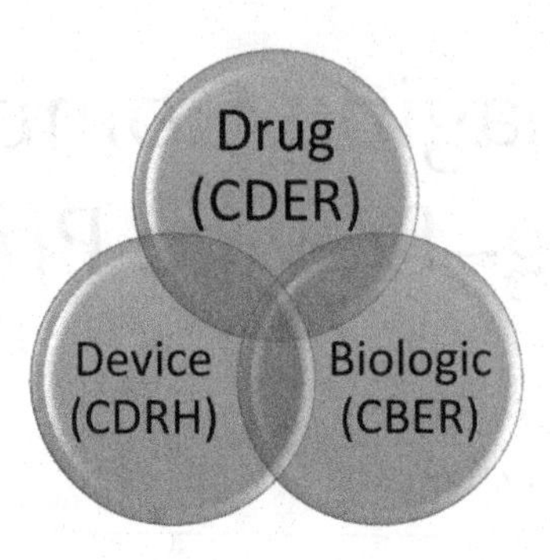

FIGURE 8.1 Venn diagram for the Office of Combination Products with the FDA center responsible for each section.

use such as a companion diagnostic that is required for safe and effective administration of a drug.

Combination products fall under the jurisdiction of the Office of Combination Products (OCP) in the FDA. Each has components that require review from a different center of the FDA (Center for Drug Evaluation and Research (CDER) and Center for Devices and Radiological Health (CDRH)), and the evidence required to support a review by these different agencies varies. Hence, the development plan for a combination product will be different and, oftentimes, more complex than a drug development plan. Creating a combination device adds levels of complexity and regulatory requirements to an already complex process of getting a product to the market (Figure 8.1).

This chapter describes the regulatory context of combination products, including a brief history of how combination products have been regulated. We will then review the regulatory framework that determines the path to approval, which, in turn, shapes the development plan. We will finish off by describing some differences a PM may notice when developing a device compared to those when developing a drug.

8.2 HISTORY OF DEVICE REGULATION

Prior to 1976, US devices did not have a dedicated regulatory authority, but instead they were regulated by drug authorities (i.e., CDER). In 1976, the Medical Device Amendments to the FD&C Act created a distinct regulatory pathway for new medical devices, including a mechanism for new investigational devices to be studied in humans (i.e., Investigational Device Exemption). In addition, the Medical Device Amendments established the three-class, risk-based classification system to all medical devices that is still used today. In 1990, the Safe Medical Devices Act further refined the regulatory pathway by introducing the 510(k) program, whereby new devices could be compared with a predicate with substantial equivalence. The act also introduced a regulation for drug–device combination products.

For many years, the FDA struggled with developing a centralized and standardized regulatory pathway for new drug–device products because they contained components regulated by more than one FDA center – CDER, CBER, and CDRH. To remedy this situation, in 2002, the FDA, as part of the Medical Device User Fee and Modernization Act, created the OCP. This paved the way for additional guidance on developing combination products.

The basic requirements for combination products were set forth in section 503(g) of the FD&C Act and 21 Code of Federal Regulations (CFR) Part 3. This established that the "drugs, devices, and biological products included in combination products are referred to as constituent parts of the combination product".[1]

In 2013, further requirements for clinical good manufacturing practices (cGMP) and post-marketing safety reporting were set forth in 21 CFR Part 4. This included clarification of which cGMP requirements apply when drugs, devices, and biological products are combined and created a "streamlined" regulatory framework when developing combination products. This was further enhanced by the publication of the Guidance for Industry and FDA Staff: Current Good Manufacturing Practice Requirements for Combination Products in 2017.[2]

8.3 DETERMINING THE REGULATORY PATH

The regulatory path determines the development pathway and plan, so it is important to decide early on which path to pursue. In addition, this critical first step governs all future interactions with the FDA. Therefore, before starting a combination product project, creating a cohesive regulatory strategy and project plan is essential.

8.3.1 Establishing the Primary Mode of Action

The primary mode of action (PMOA) of the combination product determines which FDA center will lead the review (CDER, CBER, or CDRH). Figure 8.2 outlines potential options for each center. Ultimately, the PMOA determines the regulatory pathway required, which can affect the time and cost of the overall project. Therefore, this selection requires careful consideration.

Although drug–device combination products are the most common, determining whether the PMOA is a drug or device provides several challenges. Sections 201(g) and 201(h) of the FD&C Act outline the statutory definitions of both drugs and devices. The problem is that all FDA-regulated medical products meet the definition of "drug". To be considered a "medical device", the product must also meet the more restrictive device definition of "an instrument, apparatus, implement, machine, contrivance, implant, in vitro reagent, or other similar or related article," and "not achieve its primary intended purposes through chemical action within or on the body of man or other animals" and "not dependent upon being metabolized for the achievement of its primary intended purposes".[3]

---

[1] Principles of Premarket Pathways for Combination Products: Guidance for Industry and FDA Staff.

[2] Guidance for Industry and FDA Staff: Current Good Manufacturing Practice Requirements for Combination Products.

[3] Classification of Products as Drugs and Devices & Additional Product Classification Issues: Guidance for Industry and FDA Staff.

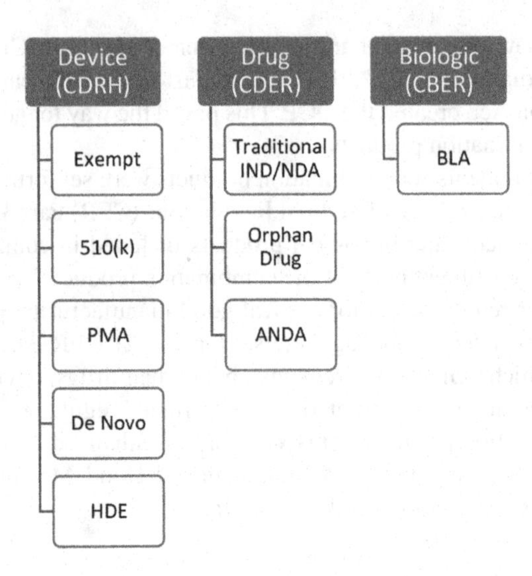

FIGURE 8.2 Potential regulatory pathways.

The FDA has a Request for Designation form to obtain a binding classification of the product. In addition, a sponsor may request a Pre-Request for Designation meeting to gather information and resolve questions with the agency before embarking on the development journey. Although the sponsor may request a specific designation for the combination product, the OCP makes the final determination with input from the relevant agency components.

8.3.2 Strategic Considerations When Picking the PMOA

The PMOA will lay the foundation for the regulatory options available. These regulatory options can significantly impact all three project constraints: time, cost, and scope. For example, the traditional drug IND/NDA pathway typically requires a longer timeline and higher costs than many of the medical device options, ultimately affecting the cost/benefit analysis. As such, developing a regulatory path that meets the requirements of your organization is a key stage-gate.

In addition, the PMOA defines the quality system requirements to produce the product. Medical device quality systems follow 21 CFR 820, and pharmaceutical quality systems follow 21 CFR 210/211. Typically, organizations select projects that match the quality system they are currently using. Additional sections to meet FDA requirements fulfill the streamlined FDA requirements, as outlined in the sections below. If the PMOA does not match the organization's quality system, the FDA does not require the organization to create a new quality system but does allow for the "streamlined" approach while realizing that they will be held to the standards of the PMOA. Irrespective of the quality system selected, documentation is required to demonstrate compliance with both 21 CFR Part 4 and the cGMPs of the selected center with the PMOA.

The PMOA selected can also have a lasting impact. It impacts both product claims for marketing the product and physician ordering of the product. In addition, if this is a new product design or manufacturing platform, it will impact not only the current product but also any follow-on products using the same design or platform.

The PMOA decision is one of the most important decisions in the product lifecycle. It requires a thorough review of the potential downstream options from the drug, device, and/or biologic standpoint. Once this path is confirmed by the FDA, changing the PMOA is difficult. Therefore, consultation with non-FDA regulatory experts from all regulatory centers involved (CBER, CDER, and CDRH) provides an informed decision. If adequate experience is not available within an organization, outside consultants should be considered.

8.4 DEVELOPING THE PROJECT PLAN

Once the PMOA has been determined, project planning begins. The project plan consists of the requirements for the FDA center responsible for the PMOA plus select additional requirements from the other center(s).

This section will focus on the aspects of the project plan that are unique to the development of combination products. The project plan for the drug development process is outlined in Chapter 11, so this section provides reference to the additional portions required for the "streamlined" regulatory approach. The regulatory options for devices are more diverse and have a much wider variety of options; therefore, this section provides a more comprehensive regulatory background along with the FDA's "streamlined" regulatory elements of the device PMOA.

8.4.1 Project Planning for Drug PMOA

Project planning for a drug PMOA closely follows the outline provided in Chapter 1. Therefore, we will not describe it in detail here. However, the OCP will expect the following additional "streamlined" regulatory elements from 21 CFR 820:

- 21 CFR 820.20- Management Responsibility
- 21 CFR 820.30- Design Controls
- 21 CFR 820.50- Purchasing Controls
- 21 CFR 820.100- Corrective and Preventative Action
- 21 CFR 820.170- Installation
- 21 CFR 820.200- Servicing

8.4.2 Project Planning for Device PMOA

When comparing these with the pathways outlined in Chapter 1 for pharmaceutical products, medical devices provide a wider variety of options. This is the result of needing to accommodate products ranging from simple bandages to complex heart valves.

A general outline of potential regulatory options is presented in Figure 8.4. This displays the diversity of options, although there are nuances and details which are not easy to capture. Hence, experts in medical device regulatory affairs are typically consulted prior to establishing the regulatory path for a particular device. Some options, such as "Exempt" and 510k, provide a lower cost and reduced time to market when compared with the drug IND/NDA pathway. Other options, such as the premarket approval (PMA), typically include clinical trials and follow a timeline analogous to the drug IND/NDA pathway.

Medical devices are separated into different categories based on the level of risk it presents to a patient. Going back to the simple bandage idea, the bandage has a low to moderate risk of failure because it is applied externally, and the only substantial risk is contamination of the wound. Therefore, the risks are infection and irritation, which are unlikely to cause severe injury or death. At the other end of the spectrum is a pacemaker, which is a high risk because it is implanted internally, has electronic parts, and is critical to sustaining life. The risks range from surgical infection to mechanical failure, any of which could potentially result in patient death.

One term used in the medical device industry which is not present in pharmaceuticals is the "predicate device". This term describes a previously marketed device that meets the requirement of "substantially equivalent" for regulatory purposes. It allows the manufacturer of a new device to demonstrate that the new device is substantially equivalent to a device that has been approved to be legally marketed in the United States. The use of a "predicate device" allows for the use of the 510(k) pathway, which typically costs far less money and takes less time than either the DeNovo or PMA pathways (Figure 8.3).

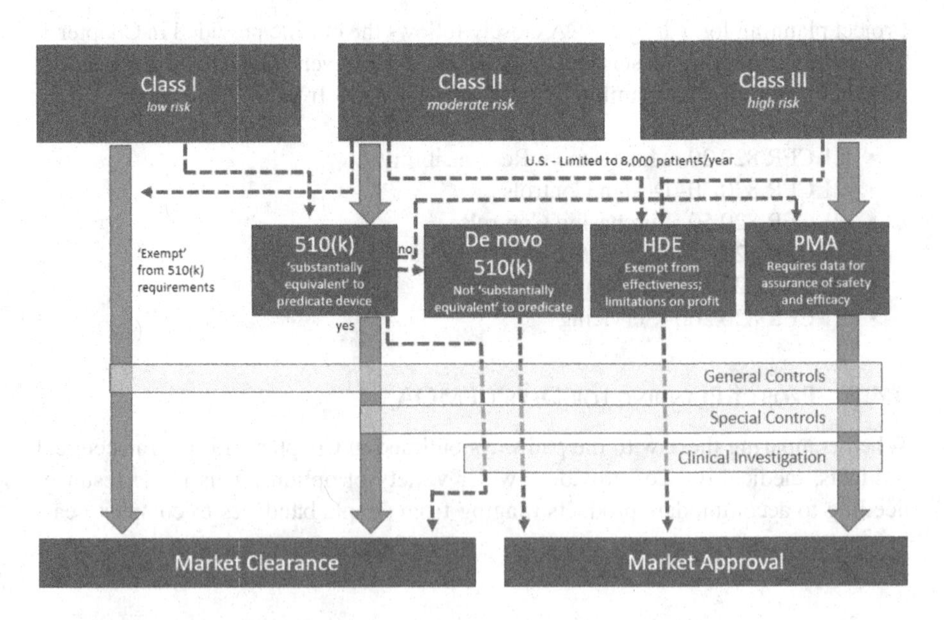

FIGURE 8.3 Diagram of potential medical device pathways.

Similar to a drug PMOA, the OCP will expect the following additional "streamlined" regulatory elements from 21 CFR 210/211:

- 21 CFR 211.84- Testing and Approval or Rejection of Components, Drug Product Containers, and Closures
- 21 CFR 211.103- Calculation of Yield
- 21 CFR 211.132- Tamper-Evident Packaging Requirements for Over-the-Counter (OTC) Human Drug Products
- 21 CFR 211.137- Expiration Dating
- 21 CFR 211.165- Testing and Release for Distribution
- 21 CFR 211.166- Stability Testing
- 21 CFR 211.167- Special Testing Requirements
- 21 CFR 211.170- Reserve Samples

8.5 DEVELOPING COMBINATION PRODUCTS

8.5.1 Development Path in Drug vs. Device

Whether you are making a drug product, a medical device, a biologic, or a combination of any of the above, the project plan needs to provide the elements to demonstrate the quality, purity, and safety of the final product. Therefore, many elements in the development path are the same among the products. These include the following elements:

- Product stability
- Toxicology
- Process validation
- Equipment validation
- Process documentation

Both pharmaceuticals and medical devices have stages of product development. The primary difference between them is that the pharmaceuticals have a very defined Phase 1 to Phase 4 clinical studies. Since medical devices have pathways that do not require clinical studies, there is no linkage to the clinical study stage. Since medical devices do not need studies to evaluate safe and effective dosage, typically no more than one clinical study is performed. The medical device stages are listed in Figure 8.4.

Throughout the development process, feedback can come from the FDA. This feedback can come from one of two places. First, they can come through application-based mechanisms through the center to which the submission will occur through the lead center for the combination product through pre-submission meetings. This approach is considered the most efficient and effective. If special information is needed, such as marketing authorization for a combination product, a Combination Product Agreement Meeting may be requested.

FIGURE 8.4 Typical device development process.

8.5.2 Considerations for Development of Companion Diagnostics (CDx)

CDx are medical devices that provide information needed for the safe and effective use of a corresponding drug or biological product. Drug developers are increasingly becoming involved in incorporating CDx into their development strategies because of the potential to identify the patients who will benefit from a targeted therapy. In some cases (e.g., oncology), a regulatory agency may require that a companion diagnostic be approved as a condition for approval of a therapy. While the benefit to patients is clear, the development path becomes more challenging for the drug product in terms of both clinical utility and regulatory approval. A PM needs to be able to create a comprehensive plan to support the co-development of the drug and the device, as shown in Figure 8.5.

CDx are class III devices, meaning they require the most stringent control to assure safety and effectiveness of the product. There are typically two ways to approach development of a companion diagnostic. The more common approach is to partner with a company that specializes in diagnostic development. The development of the drug and device then becomes an interplay between the drug developer, the device developer, and the contract research organization that is running the clinical trials. The other approach is through a laboratory developed test, which is a single-site PMA granted to a laboratory that developed the assay to support the clinical program.

8.5.3 Lifecycle Management (LCM)

As described in Chapter 18 for pharmaceutical development, medical devices go through LCM. If anything, LCM plays an even larger role in the medical device industry.

When updating pharmaceutical products, scale-up post approval changes must be followed. Depending upon the level of change required, additional studies may be required. These studies required to justify the scale-up post approval changes can be extensive and prohibitive, so changes are minimized as much as possible.

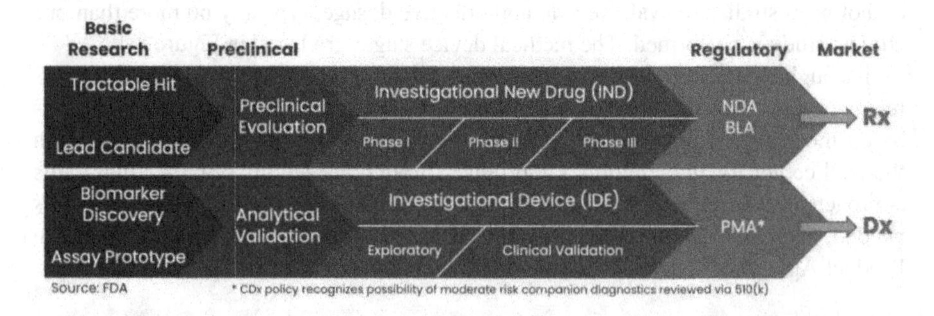

FIGURE 8.5 Diagram of companion diagnostic development process juxtaposed with drug development process.

By contrast, design changes in the medical device industry are not just expected but are the norm, especially for level 1 and 2 devices. The concept of "substantially equivalent" in the medical device regulations allows that minor changes to the design (e.g., shapes and sizes) can be made with a simple "letter to file" or a relatively simple 510(k). As a result, many medical devices have tens to hundreds of different sizes and shapes of essentially the same thing. Examples of this are all the different sizes and shapes of bandages.

8.6 DIFFERENCE IN PROJECT MANAGEMENT BETWEEN DRUG DEVELOPMENT AND DEVICE DEVELOPMENT

Several aspects of project management are different for combination products than for either medical devices or pharmaceutical products alone. This can impact what the team is doing, who is on the team, and how the team is managed. The next sections will dive deeper into these differences and what makes project managing combination products unique.

8.6.1 Quality and Risk Management Systems

Most people assume that just because all aspects of pharmaceutical products fall under the FDA that the focus of each center is the same. This is true at a high level – all centers work to ensure the safety and efficacy of products being delivered to patients. At the lower levels, differences in how this is done quickly become apparent. The differences are the result of what each center has determined to be the most critical portions of the quality systems product development and how to address the inherent risks involved with developing new products.

Early drug development of a new chemical entity involves dosing a new chemical, with the potential for unpredictable consequences. As a result, early-phase drug development focuses on determining the potential risks in animals, followed by dose-escalating studies to minimize the risk to humans and establish the risk profile. Therefore, pharmaceutical products traditionally have a result-driven project plan with formal risk analysis only performed toward the end of the project. Recently, with the advent of quality by design, more risk analysis is being performed earlier in product development.

When medical devices became regulated in May 1976, the FDA decided to take a different approach. Medical devices provide a more predictable risk profile in early stages of development. Therefore, the two types of risk analysis are initiated early in the product lifecycle – device design and process design. These lay the foundation for the regulatory path and, ultimately, the project plan.

The quality systems for drugs and medical devices also reflect these fundamental differences in the result- and risk-driven systems, respectively. Figure 8.6 identifies the primary sources of regulatory requirements used for developing quality systems and specifications within the United States. This is not an exhaustive list, and there are additional guidance and standards documents necessary for many applications (Figure 8.6).

Drugs	Medical Devices
• 21 CFR 210/211 • United States Pharmacopoeia (USP) • International Council on Harmonization (ICH)	• 21 CFR 820 • ISO Standards • Medical Devices Regulation (EU) 2017/745

FIGURE 8.6 Comparison of regulations governing quality systems for drugs and devices.

8.6.2 Use of Agile Methodologies

As mentioned in Chapter 6, a hybrid of waterfall and Agile techniques are often used in project management for pharmaceutical products. Products with a medical device PMOA are well suited for following techniques closer to the Agile methodology due to the shorter duration and the risk-driven quality system. In addition, the software development of some medical devices is well suited to the Agile philosophy.

8.6.3 Team Structure

Combination products benefit from having at least one team member who is familiar with quality systems for each of the applicable centers. For example, in a drug–device combination product with a PMOA of drug, creating a team with a team member familiar with the regulatory requirements for medical devices provides a better understanding of the "streamlined" requirements and some of the terminology differences between the two FDA centers. In companies with a large oncology portfolio, there is often a department focused on biomarker and companion diagnostic development.

8.7 SUMMARY

Drug–device combinations cover a wide range of medical products. Development of these products is often more complicated than drug development in the sense that different sets of regulatory requirements to the drug as well as the device. It is important to choose the PMOA wisely at the outset of the project because the PMOA determines the development plan and the level of evidence needed to obtain market approval. The device development process can be compared with drug development in that there are defined Phase I through V stages. In many cases, as with CDx, the development of the drug is done in parallel but independently from the device. Nonetheless, development activities must be coordinated so that availability of the products coincides at market approval.

9 How Technology Can Assist Project Managers Now and in the Future

Dave Penndorf
Planisware

CONTENTS

9.1 INTRODUCTION

Given the complexity of development projects described in Chapter 1 and the broad responsibilities for a project manager described in Chapter 2, it is no wonder that we need some help from technology to manage it all. What may have worked in the past with a spreadsheet or slide deck is no longer sufficient; project managers nowadays

DOI: 10.1201/9781003226857-11

must be able to dynamically schedule activities, analyze complex scenarios, generate data-driven reports, keep project information organized, and keep project stakeholders of varying interests informed.

The PMI PMBOK® Guide (5/e) defines a project management information system (PMIS) as "an information system consisting of the tools and techniques used to gather, integrate, and disseminate the outputs of project management processes. It is used to support all aspects of the project from initiating through closing, and can include both manual and automated systems." Simplifying this definition, a PMIS is the set of tools and processes used to manage projects. For project managers, the PMIS becomes the system of record for project information, and as owners of project information, it is paramount to know your PMIS well.

In this chapter, I will discuss various types of tools which make up a PMIS. The most obvious is a project management tool, but resource management, budget management, project information, and communication tools all comprise a PMIS. This chapter and the next one complement each other. This chapter describes the purpose and utility of various tools that PMs use within the PMIS. The next one explores the key features to consider when selecting a project management solution and best practices for implementing a project management solution in your organization.

9.2 PROJECT MANAGEMENT TOOLS

The past decade has seen an explosion of tools to support project management at various levels of sophistication and maturity. The introduction of simple task management solutions has enabled the casual project manager to keep track of activities without requiring formal PM training. More traditional and robust project management solutions continue to add features and functionality to support increasing demands on project managers. And broader project portfolio management (PPM) systems have been introduced that cover not only project execution but also information to support decisions on intake, selection, and value realization of projects.

Figure 9.1 gives some examples of these basic, traditional, and advanced project management solutions.

9.2.1 Task Management Software

The purpose of task management software is to enhance the collaborative environment on a project via a collection of the work that needs to be accomplished. Task management tools are intended to democratize project work, meaning the project team members can interact with the tools to view their work, send communications about their tasks, and update progress toward completing the task. Usually, these tools are considered very easy to use and require minimal specialized training. Task management software is typically better for shorter term task execution, as opposed to longer term forecasting or strategic decision-making that requires a more robust solution.

Historically, task management software has not had a place in the life science industry because of the complexity involved in such projects, but improvements in technology have opened the door for their involvement in the PMIS space. Some

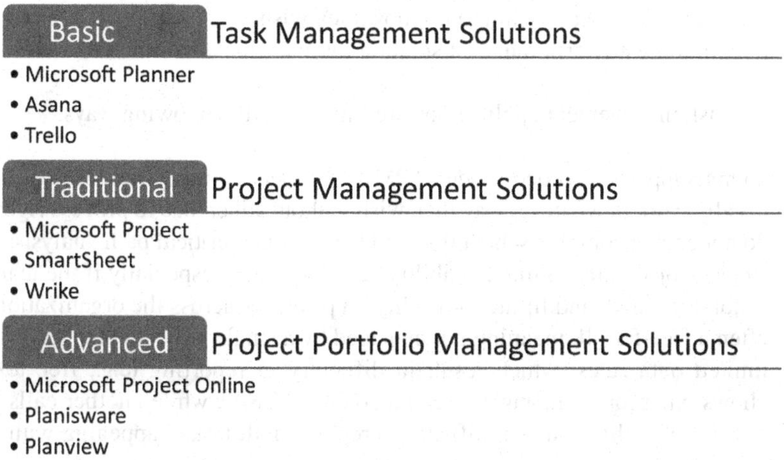

FIGURE 9.1 Common project management solutions grouped into basic (Task Management), Traditional (Project Management), and Advanced (Project Portfolio Management).

organizations are implementing task management to help project managers coordinate, disseminate, and monitor the work that needs to be done.

Most task management solutions come with a few ways to visualize work such as

- list view,
- calendar view,
- Kanban or card view, and
- simple Gantt view.

To be sure, even with their advances, such tools remain insufficient for development project managers to do their job. For instance, I have seen task management tools fit in as follows: a project manager still uses a robust project-scheduling tool with subtasks, dependencies, Gantt views, and resource loads as their personal project management tool, and then duplicates the basic task information in the task management tool just in time when a task needs to start being worked on. In this way, the project team only interfaces with the intuitive task management tool, while the project manager still has the dynamic scheduling capabilities of a project planning tool.

Another use case may be for managing subsections of a development plan with a dedicated set of team members who need to execute a higher number of granular tasks than the development project manager cares to maintain, for instance, site initiation, or coordinating all sections of writing and finalizing the Common Technical Document.

Task management applications should have the following features:

- easy to use, with minimal training required;
- foster transparency, which aligns team members on who is doing what;
- may support various planning methodologies (Kanban, Agile, hybrid planning);

- often flexible for each team to use how they wish;
- integration with corporate chat solution for convenient communications.

However, task management applications are limited in the following ways:

- do not support the process side of PMIS – project managers and teams can usually work in whatever way they wish without adherence to process rigor;
- do not enable complex scheduling and have limited critical path analysis;
- single project only: limited visibility across projects, especially if the team is not dedicated, and limited visibility on priorities across the organization;
- often a free-for-all, meaning not managed by a single project manager
- limited data rules, which result in difficulty in reporting (e.g., free text allows one project manager to call a country "USA", while another calls it "America" – this makes it difficult to report on all tasks happening within this country).

9.2.2 Project Management Software

Project management software is more robust and powerful than task management software, and it is the "go-to" tool for professional project managers. Project management software is mostly focused on scheduling. Project management solutions usually adhere to standards for scheduling such as the Project Management Institute (PMI), although there are increasingly more tools that allow flexible, unstructured planning such as used on Agile projects. They may support phase-gate processes as well by allowing certain milestones to be defined that trigger "go/no-go" decision points. Most solutions can integrate resource and cost planning into the schedule, although typically not as a strength and often with limitations in capacity management or specific finance requirements.

With the focus on scheduling, a project management tool helps project managers organize and manage work by allowing them to create a complete list of activities that need to be completed to finish the project. These activities can be connected with dependencies that allow the project manager to change the timing or duration of one task and use the scheduling engine to automatically change the timing of all downstream tasks. In addition, the connected tasks allow the project manager to analyze critical path activities that need special attention on the project. Costs and resources can then be applied to these activities or their parent work packages, summarized, and time-phased for comprehensive analysis of project metrics.

The predominant tool for communicating a project schedule is the Gantt chart (named for Henry Gantt, which is why Gantt is always capitalized), and every good project management solution should have the capability to create one. A Gantt chart provides visibility to the full scope of work, overlapping activities that may cause resource availability bottlenecks and dependencies between tasks that may require additional attention. This visibility is particularly useful in driving alignment within the project team, by showing clearly where and to whom each task hand-off is occurring. As shown in the example in Figure 9.2, the project's activities are plotted over time and connected with arrows to indicate interdependencies.

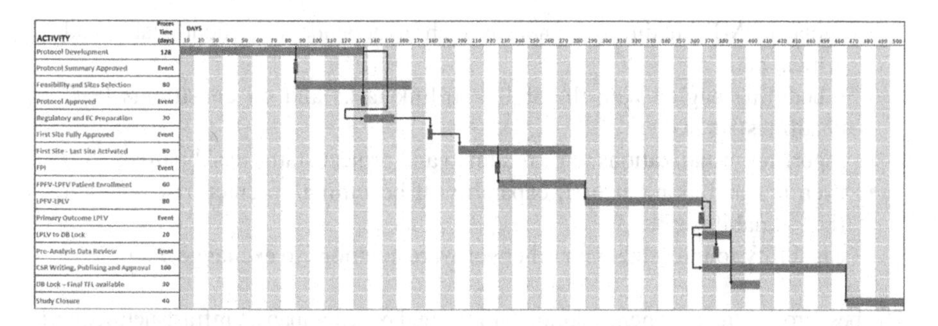

FIGURE 9.2 Example of Gantt chart

The project manager can use a Gantt to share all or pieces of the plan to any stakeholder, including leadership, project team members, functional area heads, contractors, and partners. However, complete Gantt charts often include more detail than is easily digestible, even by skilled project managers, let alone by stakeholders. To narrow down the number of tasks into digestible chunks, the project management solution should support sorting, filtering, and formatting to aid the project manager for effective analysis and reporting. Visualizations can be utilized to enhance analysis and communication, using either stand-alone tools like OnePager or Milestones Pro, or via PPM tools which have a roadmap or timeline capability built-in like Planisware or Microsoft.

Project management software will almost always allow for project templates, which help a project manager not only as a time-saver when creating a new project plan but also as a checklist of activities that need to be completed for a certain type of project. Templates should certainly be used for projects that are run repeatedly, such as a clinical trial, but are applicable starting points even for novel development paradigms. Templates can come pre-loaded with benchmark task durations, resource requirements, and costs, all of which allow the project manager and team to assess the needs of the project compared with a "standard" project. We will discuss an "enhanced" way to pre-load information into project plans in Section 9.7.2.3 Algorithms.

Project management software can be *stand-alone*, in which each project plan is siloed and does not share data with other projects, or *enterprise*, in which the projects all reside on the same system and can share data. Example data to be shared between projects include inter-project dependencies (e.g., finish-to-start links), working or holiday calendars, pool of resources, metadata (e.g., lists of dropdowns, like the aforementioned example of country names), and collating data from the projects to do cross-project reporting.

Many add-ons exist to enhance project management software, for instance, adding resource capacity planning or strategic portfolio planning capabilities on top of best-of-breed scheduling engines. Such an approach enables project management software to approximate PPM software (see the next section) although at the cost of maintaining separate solutions and trust in their interoperability.

When evaluating a project management solution, consider the following:

- support PMI scheduling standards such as dependencies, float, slack, and critical path;
- scheduling engine with the ability to link tasks and set constraints (e.g., deadlines) on tasks;
- reporting visualizations (e.g., roadmap and project status dashboards);
- even if stand-alone at first, capability of becoming an enterprise solution (i.e., scalability);
- capability to integrate with other software solutions to exchange data.

The best project management solution will depend on a company's infrastructure, portfolio size, and number of users. In general, medium- to large-size companies will need more sophisticated enterprise-level software, whereas smaller companies can usually get by with smaller, less complex applications. Even smaller companies are seeing the benefits of a robust project management solution, and I have seen companies as small as 100 employees with no clinical trials implement a robust solution to lay the groundwork for the growth of employees and project complexity that will come as soon as their pipeline moves into the clinic.

See Chapter 10 for more discussion on implementing a project management tool.

9.2.3 PPM Software

PPM software extends beyond the focus of only the project manager. In PPM, the focus broadens to include how a project fits with the rest of the organization to be successful. Resource, cost, strategic alignment, and reporting are often key components to PPM software for an organization. An oft-cited term in the PPM software sphere is to provide a "single source of truth" to the underlying project data – in other words, an organization relies on the data contained within the software for its decision-making without having to seek or vet the data.

A PPM tool need not shirk on the scheduling features: in other words, the project management aspect can be as rigorous as a project management software, although some PPM solutions do intentionally limit the depth of project management protocols supported as a trade-off with ease of use.

PPM software will be "enterprise" by nature. Sharing data is a key value-add and thus an important component to setting up a PPM solution. The data model and data dictionary must be defined – for instance, which calendars to use, which dropdowns to define, and with which possible values. Even if the scope of a PPM software implementation is scheduling-only (i.e., no resource or cost planning), such data model decisions typically require input beyond the project management group, for example from the Finance team.

A project manager may ask, "what is in it for me" to utilize an integrated PPM software solution, instead of a dedicated project management solution. Consider the following real-world use cases I have observed streamlined through the use of PPM software:

- With proper integrated resource planning, aligned with the functional teams and budgeted by governance teams, a project is less likely to be impacted by a resource bottleneck.

- With alignment of direct expenses to a schedule, a project manager has bottom-up evidence to justify seeking a budget extension when a project plan changes.
- Instead of various stakeholders requesting project information from a project manager at all hours of the day, the stakeholders are able to directly access the information themselves.
- In evaluating risk mitigation options, a project manager can conduct what-if scenarios efficiently, without having to pass data offline back-and-forth with various integrated teams.
- Reduced manual steps through software support (see section 9.7.2 below).

Each of the above examples may have required additional up-front effort from the project manager in planning but resulted in risk mitigation, reduced long-term effort, and improved project outcomes.

PPM software should have the following features:

- all features listed above in task management and project management software;
- flexible scheduling capabilities for different kinds of projects, including full PMI support, Agile, and hybrid planning;
- ability to extend and fully control the data model and data dictionary;
- built-in reporting visualizations and capability to integrate with dedicated business intelligence tools;
- extensive integration capabilities as the PPM software will likely need to exchange data with various other systems in the organization's ecosystem (see The Broader Ecosystem: IT Dataflow in Section 9.5).

9.3 RESOURCE MANAGEMENT TOOLS

As described in Chapter 4, resource management is among the most essential, yet inadequately performed, aspects of business management. The aim of resource management is to ensure the organization has the right resources available at the right time to do work. Resource management involves capacity planning at a high level and named resource assignment at a granular level. Often, the tools for resource management are integrated into the project planning software, but there are also stand-alone solutions or bespoke home-grown solutions that may be appropriate for biopharma.

Technology tools can assist project managers with two aspects of resource management: (a) projecting how many resources will be needed to support the organization work (i.e., resource capacity planning), and (b) obtaining resources for a given project (i.e., resource assignment). The tools can range from simple spreadsheets that add up full-time equivalents (FTEs) in columns and rows to very complex systems that involve algorithms and programmatic matching features. At very small organizations, a simple spreadsheet may suffice for estimating resource demand. However, as the organization grows and the number of roles expands, more sophisticated tools are quickly required.

9.4 BUDGETING TOOLS

In some industries, budget planning is a key part of a project manager's remit. Indeed, according to PMI best practice, the project cost is a rollup of the costs for each work package in the project plan. As work is completed on each work package, the actual costs are entered and easily compared to each work package. While some biopharma organizations are able to imbed their project management solution with their finance processes, in most, the FP&A (Financial Planning & Analysis) function maintains the budget separately. This separation of project work and the costs associated with the work creates challenges for project managers to estimate if spend is on track or not.

To further complicate the tool landscape, when FP&A manages the budget and actuals in a separate system, the work packages often do not match. For example, the project plan may have separate line items for three drug–drug interaction studies, a renal impairment study, and a hepatic impairment study, but the FP&A budget has a single line for "clinical pharmacology studies". It is therefore not possible to estimate if the spend on each of the studies is on track.

9.5 TOOLS TO MAINTAIN PROJECT INFORMATION

There is a lot of project information generated over the life of a drug development project. Any project manager who has taken over a program mid-flight knows that a haphazard file organization structure can make it difficult to figure out the history of the project. A good project manager, with the help of the Project Management Office (PMO), will figure out a way to organize project information that works at the time of inception, lasts long into the future as team members change, and is scalable as the project grows.

A project manager is responsible for creating an environment where team members can get work done. When it comes to working on cross-functional documents, team members need to be able to work on the same document at the same time to avoid redundancy or version control issues. The advent of collaboration platforms like SharePoint, Box, Egnyte, and Google Drive has eliminated the need for people to either work on documents in sequence or to work on their piece and then the project manager has to stitch it together. Gone are the days of emailing documents around and having to deal with multiple versions coming from every author. Now everyone can be pointed to the same document, so they can work on their content simultaneously.

However, the project manager must sometimes remind people that changing document creation from an individual sport to a team sport requires new ways of working. For example, do not check out documents and forget to check them back in, do not overwrite someone else's edits (use comments instead), and do not save the file with a new name.

In addition to project documents, there are also pieces of key project information such as goals, risks, issues, decisions, actions, and lessons learned that should be captured. Part 3 of this book provides a detailed description of the GRIDALL methodology and how it can be used to capture project information in a simple, comprehensive, connected, and feedback-looped framework.

Transparency is one of the keys to success with GRIDALL. Using intentional folder structure and document management provides the ability for team members to successfully search for any aspect of the GRIDALL methodology that pertains to them. Even better, with a robust PMIS solution, all of this information can be tracked within one tool so that team members can review their upcoming activities at the same time they review or even participate in the tracking of any of this key project information.

9.6 TOOLS TO COMMUNICATE WITH TEAM MEMBERS

The adage that 90% of a project manager's time is spent communicating is still true and relevant in today's hybrid work world. In the past few decades, communication tools have evolved to enhance the ability to move project information quickly and effectively throughout the project team and to other stakeholders. This communication can take several forms in the project environment:

- direct discourse between two or more people using meetings or videoconferencing tools,
- electronic correspondence via email or chat, and
- broadcasts of information via reports and dashboards.

9.6.1 Meetings and Videoconferences

Sometimes a good old conversation is the best way to share information, although with an increasingly remote workforce; in-person interactions are becoming less convenient. Enter videoconferencing tools. With the click of a button, you can be on a voice- and video-sharing call with a colleague or group of colleagues. These interactions do not even have to be scheduled anymore because now you can see with an icon whether a person is available to talk. Video calls have become so easy and convenient nowadays that people will often make a call, rather than get in an elevator to see a colleague on another floor.

As a remote project manager, I have found video calling tools to be sometimes more effective than trying to gather people physically. For example, I might be discussing a topic with one colleague, and we realize that we need some input from another colleague. I can then dial that other colleague into the call within seconds, rather than walking around trying to find the person.

It is important for a project manager to know the limits of video calls. During the COVID pandemic, people started to feel the strain of the consistent and relentless scheduling of meetings. This so-called "Zoom fatigue" left people more exhausted at the end of a day than a day filled with in-person meetings that required people to walk between conference rooms and get that chance to change their scenery every so often. Furthermore, more research is emerging that reveals the stress people feel by seeing themselves on camera, and so the choice for "video optional" meetings is becoming more common.

Project managers may also want to adapt their meeting norms in a hybrid environment. With in-person meetings, I used to get offended when an attendee was multitasking during the meeting mostly because typing away on a computer is distracting for other attendees. With videoconferences, I see an opportunity for people to multitask without distracting others, and I can pull them into the conversation when they are needed. To me, that is an efficient use of time.

9.6.2 Email and Chat

Emails are a perennial communication tool for one-way, asynchronous correspondence. Email software has improved over the years. For example, you can now create an email and schedule it to be sent at a certain time – helpful for those who work in different time zones or in the middle of the night and do not want to load people's inboxes before they start the day. Using the "Read/Unread" feature allows you to read an email and then mark it unread if you want to come back to it. You can also move emails that require action to your to-do list for future action. There are even attempts to unleash natural language processing and artificial intelligence (AI) to your inbox and identify when you have committed to doing something or someone is asking you to do something, creating a digest of action items as a reminder.

Instant messaging, or chats, have also found a place in business for quick correspondence between 2 people or a small number of people. Some of the current chat tools include Microsoft Teams, Slack, Cisco Jabber, and Google gChat. For larger audiences, I have found chat threads to get too complicated, and it is too easy to lose the conversation. One useful case for large chat groups is as a way to broadcast announcements in a one-way push of information. In this way, the chat thread becomes a news feed, rather than a dialogue.

One more cautionary tale about chat tools: they seem to be popping up all over the place, including in project management applications. However, using too many communication tools can make it difficult for people to keep up with all the information. My advice is that unless it is the main communication channel for your organization, it is better to avoid using the chat functionality built into project management applications. As a best practice, the maximum number of communication channels is 2 – email is perennial, and chat is now commonly used, and that is all people can handle.

9.6.3 Reports and Dashboards

Anytime project information is shared with stakeholders, it can be considered a "report", even if it is just text in an email. Reports and dashboards refer to consistent or reusable layouts and formats and are common ways to share project information with stakeholders. Reports can be static, meaning their data do not change, or dynamic, meaning the data refresh automatically. Slide deck reports, a perpetual favorite among executives, are an example of a static report because they are easy to consume when created in a consistent layout and format. However, there are many pitfalls with static reports and dashboards; some are given as follows:

- Reports should be easy to generate and easy to understand, but often project managers spend a lot of time creating reports that end up being confusing and difficult to interpret. (See Section 9.7.2 on ways to automate reporting using robotic process automation (RPA).)
- A common database of information is available that will answer all the various stakeholders' queries, but often a project manager will have to slice and dice the data to meet the needs of a particular stakeholder group.
- Reports need to be accurate and timely in order to tell your project's story, but manually created reports are prone to error and quickly become out of date and worthless.
- A stakeholder receiving a report may want further information to provide context, to drill down into details, or to filter data in a different way than were presented; a static report requires going back to the project manager to iterate, sometimes multiple times.

Dynamic reporting and dashboarding tools can help the project manager to save time and increase accuracy of data. Dynamic reports can be created one time and then be refreshed with new data with a few clicks of a button. Interactive dashboards connected to automatically refreshed dataset allow stakeholders to customize the data display to meet their needs, and they can be confident in their conclusions because the data are sourced in real time. When the data are fully integrated, stakeholders can drill down to see details underlying the report or to provide context for the data in the report. Common stand-alone dashboarding tools include Microsoft Power BI, Tableau, and Spotfire. Many more robust project management solutions have their own dashboarding tools, such as Planisware, Microsoft Project Online, and Planview.

A common report in biopharma is an asset-level report that shows all the clinical development plans (indications) for that asset. This type of report allows senior leaders to understand the timing of catalyst events such as data readouts that enable go/no-go decisions and regulatory filings that trigger launch planning activities. An example of such a report (generated from OnePager Pro) is given in Figure 9.3.

Another common report type for a small biopharma company is a pipeline report, which shows all the activities for all assets in the pipeline. This type of "book of work" report shows not only the milestones that support decision-making and potential funding opportunities but also the overlap of activities that may require common resources and thus may strain the organizations' workforce. At larger companies, the entire pipeline will typically not fit on a single page, so they are often broken down into therapeutic areas (e.g., oncology pipeline, or even oncology early development pipeline) (Figure 9.4).

9.7 HOW ADVANCED TECHNOLOGIES WILL BENEFIT PROJECT MANAGERS

Current project management tools require appreciable time to keep updated and to share information with others. The future of project management tools aims to automate many of the manual processes project managers do today and to enhance the

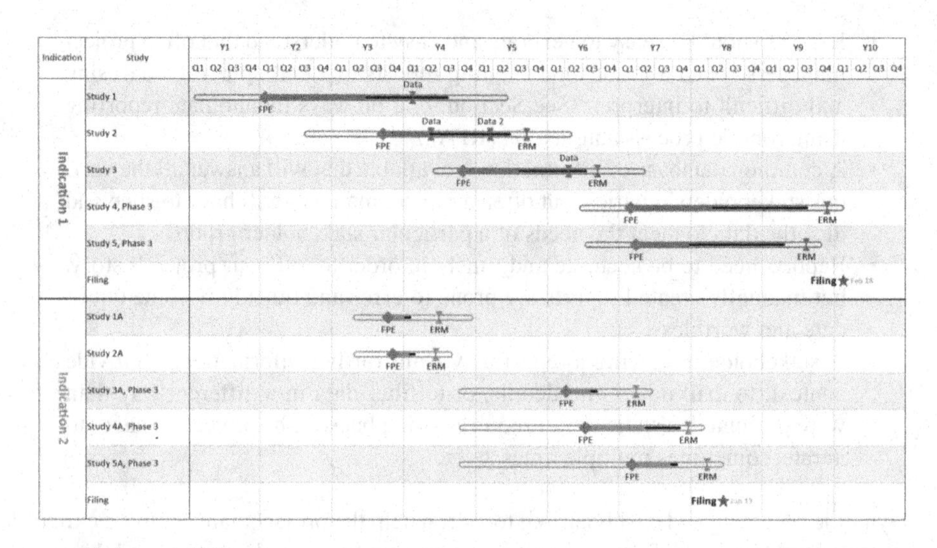

FIGURE 9.3 Example of asset-level report generated from OnePager by Chronicle Graphics.

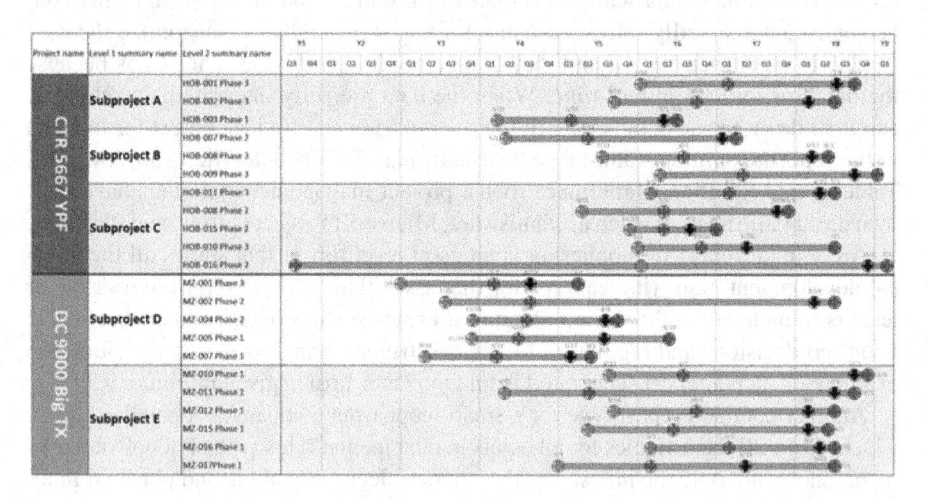

FIGURE 9.4 Example of book of work report generated from OnePager by Chronicle Graphics.

accuracy and predictability of project schedules, resource plans, and cost estimates. AI and RPA are two such tools that have potential to reduce project managers' manual duties, thereby freeing them to do higher level activities. In this section, we will explore how these advanced technologies might benefit project managers in the coming years.

9.7.1 Artificial Intelligence

Seemingly every academic discipline is investing in AI to accelerate change, break through barriers, and for associated companies to differentiate themselves. Drug development is definitely included, with quotes like "Machines will soon be the best inventors of every biotechnology..." by inventor, startup founder and Flagship Pioneering partner Geoff von Maltzahn in a May 11, 2022 tweet. The discipline of project management is no different – what are the goals of AI within project management and are they achievable?

9.7.1.1 Predictive Estimation

An oft-cited goal of project management is to achieve a reliable prediction for key variables of a project, such as task durations, resource requirements, or costs. While analogous and parametric estimation are useful tools for project managers currently, these tools are limited by the time it takes to find useful analogs and identify relevant parameters. The application of AI holds the potential for more accurate prediction through its capability to search through vast amounts of data.

For example, let us consider some high-level tasks of a typical clinical trial: startup, execution, and closeout. A project manager might be able to estimate the durations of these tasks through past experience and by looking at a few similar trials (matching the study phase [1, 2, or 3], indication, number of patients, and number of sites). However, these parameters do not capture all the nuances of the trial, which is a seamless Phase 1/2 with the escalation phase bring a modified 3+3 design and the Phase 2 portion commencing immediately after the predicted human efficacious dose is cleared. Furthermore, the indication is not simply a disease state but selected for patients with a particular genetic mutation.

A logical step to using AI in this example is to have the technology review the project manager's estimates and highlight those that are unrealistic. See below for a simplistic example – the AI system is alerting the project manager to review the estimate of the execution task. The project manager can then raise the concern to the team as a risk and assess the estimation further.

Task	Duration [Estimated by project manager]	Output of AI Tool: Confidence Indicator
Startup	6 months	Very likely for this study type
Execution	12 months	Unrealistic, given the study design and patient population
Closeout	4 months	Probable, given the number of subjects, therapeutic area

Although the appeal of AI-driven predictive estimation for project managers is exciting, in my experience to date, there are significant challenges to the successful application of AI to biopharma projects. The traditional (and admittedly simplistic) model of AI is of machine learning: using a training set of data to then predict an outcome with a novel stimulus. For example, autonomous driving cars traverse roadways with human drivers to collect "training" data before being permitted to actually drive autonomously. Estimates predict at least 11 billion miles of on-the-road training will be necessary to match human abilities. The challenge for machine learning to produce productive outcomes comes down to both the quantity and the quality of training data.

For biopharma companies wishing to apply AI to their project management practice, the quantity and quality of the training data impose considerable limitations. While some "industry standard" data can be gleaned from publicly available sources such as clinicaltrials.gov, rich training datasets are typically limited to the data coming from the organization's own past. Therefore, the training dataset has the following limitations:

- Quantity:
  - Considering the clinical trial dataset, even the largest biopharma companies have access to historical data of only a few hundred or thousand studies.
  - Processes and ways of working change over time, so the datasets often have to be culled to a narrow specific time frame, which further limits the number of applicable studies.
- Quality:
  - Change in the organizational structure may render some historical data meaningless.
  - Drug development programs that did not complete, due to toxicity or intolerability, lack of efficacy, or strategic business decision, often are not updated (imagine if your project is being "killed", updating the actuals becomes less of a priority).
  - Time tracking data are often not trusted due in no small part to poor compliance.
  - Human factors in project management result in data that are not consistent and thus not able to be mined by AI tools (e.g., one project manager may use "FPI" while another uses "FPFV" to mean the same "start-of-enrollment" milestone).

As a result of these limitations, I have yet to see a biopharma organization be successful in implementing AI-driven predictive estimation. That being said, progress requires investment in effort. Organizations are actively working to set themselves up for success in the future by expanding the quantity of data they store and grooming data quality in order to be able to leverage such AI benefits in the future.

9.7.1.2 Risk Identification

Early identification of possible risks would enable a project manager to better assess mitigation options. For instance, data scientists are tracking subject dropout rates to identify patterns through subject enrollment in order to mitigate additional enrollment risk. When looking at the cross-functional nature of a drug development plan, other patterns could be ascertained through the communications between teams.

For example, if the CMC team is constantly asking if FPI is on time, maybe it is a hint the initial drug supplies would not be available on time. A system that could therefore monitor all project communication – from email, to chat, to webconference meetings, to status reports – could use natural language processing together with traditional AI to reveal risks.

Again, the caveats will be present. To execute on such promised AI, training data would be needed on historical project communication, which is definitely lacking. Furthermore, natural language processing, although capable of great transcription, is not yet mature enough in the field of natural language understanding to assess the nuance in language that produces the insights on risk detection. Therefore, this value-add of AI for the project manager is likely even further in future than predictive estimation.

9.7.2 Robotic Process Automation

Taking over administrative or repetitive tasks for project managers can reduce non-value-added burden, freeing the project manager to focus on higher level, complex activities such as coordination and decision-making. Such tasks are termed RPA, and they can take many different forms, including team communication, algorithms, data quality controls, and even project setup.

9.7.2.1 Automated Reports

I have heard from many in the project management community that creating reports, such as monthly status dashboards or project roadmaps, is tedious and time-consuming. A good project management solution will take advantage of RPA to generate repeatable reports automatically at scheduled times. Consider a team of 20 project managers who can save 5 hours a month with automated reporting: at $200/hour, that is, $240,000/year, of savings by implementing RPA-driven reporting processes. Automated reporting of static reports can not only save a project manager time but also ensure better data quality because it avoids rekeying data into a manual report.

RPA-driven automated reports require a few special considerations, as follows:

- Standardization in process such that a majority of reports requested are repeatable. This may require harmonization on KPIs, terminology, or PPM maturity among stakeholders, an outcome whose effort should not be underestimated.
- Flexibility in the solution because the aforementioned process is subject to change and because ad hoc needs will always arise. Examples of this include (a) the ability to include or exclude certain activities from a timeline roadmap, (b) selection of which set of baseline (or snapshot) data to include in the report, and (c) variation in formatting, layout, and filtering to meet different queries.

- Ensure the report builder is available directly on the project management solution, without requiring a data intermediary (such as a data warehouse) to maximize efficiency for the project manager.

9.7.2.2 Chatbots

Project managers spend a lot of time satisfying one-off requests for common project information. Project team members and stakeholders regularly ask the project manager for data, requiring attention from the project manager to quickly address the question. Examples may include the following:

- "What is the date of the next decision point?"
- "What is the budget for the next phase?"
- "What resources are assigned to the next work package?"

The potential for chatbots to automate the responses to these queries will free up the project manager to do other things. On the flipside, the project manager regularly seeks input from various team members, for instance, seeking percent complete of a deliverable or number of subjects enrolled in a particular study. In either case, if the organization's corporate chat solution is integrated directly into the project management software system, the manual effort of the project manager decreases.

PPM solutions such as Planisware are beginning to integrate with corporate chat such as Microsoft Teams to bring this vision to reality. Utilizing natural language processing, team members can directly query the project themselves to get data they need to understand and execute their tasks. In addition, a project manager can send out queries, tracked and automated with simple embedded forms into the team members' chat windows. As these technologies mature, I envision the project team of the future utilizing their PMIS and chat tools dynamically to save time and energy.

9.7.2.3 Algorithms

Algorithms, also known as parametric estimation, are widely used across the biggest pharma organizations in their PPM system to alleviate the effort of the project manager in an intelligent automated way. With a well-tuned algorithm system, the role of the project manager is to define the work breakdown structure and enter into the system some key metadata (or "drivers") that allow a predetermined algorithm to calculate an output. Most frequently used for resource or cost estimates, it can also be used to drive task duration. A classic example: to estimate the number of site monitors needed across the life of a study, enter the number of sites, number of subjects, and therapeutic area of a study.

Algorithms are in fact merely a set of assumptions and equations, sometimes super simple, sometimes mathematically or logically complex. They become a form of RPA when they are applied to new projects (e.g., a new clinical trial) or revised projects (e.g., adding more sites to a trial) because they can be applied automatically, without manual entry from each function. The only thing that needs agreement is the project's drivers and characteristics that determine which algorithm would be applied. After that, the project manager clicks a few buttons to apply the functions' inputs.

While algorithms can be run in a separate tool from the schedule (even Excel), the project manager gains efficiency by integrating them into the PPM system directly so that new projects can be quickly created and scenarios can be generated. For instance, a project manager prepares for a team meeting with multiple what-if scenarios run ahead of time to demonstrate to the team the effect of a proposed change. The benefits to the project manager include the following:

- greatly reduced cycle times for planning cycles;
- efficient what-if modeling of major changes without requiring input from each stakeholder;
- alignment across the organization on the central assumptions driving resources, costs, or schedule.

It is important for relevant stakeholders to be involved in the creating and adjustment of algorithms. The assumptions and equations are best defined by the relevant stakeholders, for example, functional area heads for resource algorithms, FP&A for cost algorithms, and the PMO for duration algorithms. I recommend the project management community be tuned in to this process because the project management template may be affected, for instance, with resourcing-specific tasks (see Chapter 10).

9.7.2.4 Monte Carlo

Monte Carlo, an algorithm method used for decades, is a type of RPA in which scenarios are run multiple times (typically many thousands of times) to assess the outputs statistically for getting insights into the range of possible outcomes. Although more often used at the portfolio level of strategic investment analysis, Monte Carlo analysis is used by project managers with mature processes to assess risk variables to their project plan.

The most common project management use case is to analyze critical path activities to understand the likelihood of achieving the target. For example, Monte Carlo simulations can be applied to the enrollment projections for a clinical trial to determine the likelihood of finishing with a defined period.[1] Let us say a study team estimates that it will take 12 months to enroll a particular trial. The estimate assumes that all sites will open to enrollment at the same time, that no sites will fail to enroll, that patients will enroll on a regular basis, and that there will be no pauses in enrollment during the trial. However, we know these assumptions to be unrealistic. Therefore, by applying Monte Carlo simulations that allow these assumptions to vary, a project manager learns that the single best guess for enrollment duration is 16 months, with a 95% chance that enrollment will be between 11 and 25 months. Thus, the project manager notes the risk that enrollment may take 4 months longer than currently

[1] https://www.cytel.com/hubfs/Patient_Enrollment_White_Paper_Final.pdf#:~:text=Our%20proposed%20technique%2C%20known%20as%20Monte%20Carlo%20Simulation%2C,The%20first%20reality%20is%20that%20enrollment%20is%20nonlinear.

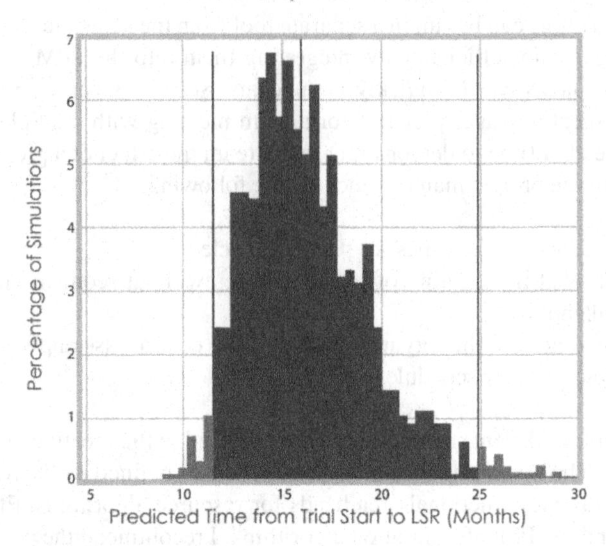

FIGURE 9.5 Example Monte Carlo simulation of predicted time from trial start to last subject recruited (LSR).

planned, and the project manager may prepare a scenario to account for the 16-month duration (Figure 9.5).

The challenge for meaningful Monte Carlo analysis is in the accuracy of the risk estimates. Project managers must predict various ranges of scenarios (typically pessimistic, likely, and opportunistic) and the likelihood of each scenario – this is the type of probabilistic estimates that human beings are notoriously bad at. I have heard that Monte Carlo analysis used to provide project manager's "CYA" in their estimate, but not actually drive better organizational decision-making. And in reality, it is the interplay between cross-functional efforts that drive variability in scheduling – for instance, if CMC delays on one program push out the availability of clinical drug supply for another.

From a software solution perspective, only a few tools on the market (including Excel) have embedded Monte Carlo analysis. If an organization's PPM tool does not, then users will have to find a way to rekey information, and the assumptions driving those variances will be dissociated from the rest of the project information. Therefore, if Monte Carlo is a tool used by an organization's project managers, I recommend seeking a tool that supports its calculations natively.

9.8 SUMMARY

With the rise of the project management discipline in many industries, tools to support project managers have increased in number and maturity. There are tools to support task management, project management, and portfolio management. There are tools to support resource management and budgeting, often integrated into project and portfolio management solutions. There are tools to manage the abundance of project information that is generated throughout long project lifecycles. And there are tools to assist project managers in communicating both to enhance the project team's dynamics and to allow rapid sharing of project information.

The future of project management will benefit from advanced technologies such as AI and RPA in that many of the manual, repetitive tasks the project managers have to do today will be replaced by my software. This has the potential to free up project managers to do higher level activities but also will require project managers to be familiar and comfortable with how the new technologies work.

10 Assessing, Selecting, and Implementing a Project Management Information System

Dave Penndorf
Planisware

CONTENTS

DOI: 10.1201/9781003226857-12

10.1 INTRODUCTION

In the last chapter, we reviewed various tools that support the project manager. Of key interest to project managers is the project management solution. In this chapter, we will explore considerations for selecting a project management solution and then describe a best-practice process for implementing a project management solution within an organization. The two chapters are complimentary and designed to be read together.

This chapter will address three types of project management solutions, briefly summarized:

- **Task management**: simple tools to disseminate the list of short-term work to be done to team members and receive updates on their status of each task
- **Project management solutions**: stand-alone tool dedicated to the project manager to schedule and coordinate their one project
- **Project portfolio management (PPM) solutions**: a project management solution with integrated additional process support to create a single source of truth (from integrating functions, to resource and cost management) used across divisions and usually supported by the project management office (PMO)

In Section 10.2, I will look at when one of the above three types of project management solutions may be appropriate for an organization. In Section 10.3, I will address implementing a project management solution within a biopharma organization and various process considerations for the project management community. In Section 10.4, I will look at integrating project management processes with resource management, while in Section 10.5, I address integrating the project management solution with other enterprise software systems and processes. Finally in Section 10.6, I will cover change management, the key to success in implementing any new tool.

10.2 NEEDS ASSESSMENT WHEN CONSIDERING A PROJECT MANAGEMENT SOLUTION

Ultimately, the right solution for an organization depends on the needs the organization must solve. The organization must be transparent about its challenges, prioritize them, and assess which are addressable by a project management solution. Below are five real-world assessments of organizations I worked with, and the best direction for them in implementing a project management solution. With these illustrations, I aim to give a type of rubric to find some similarities for your own organization and guide you on what direction to take with a tool selection.

10.2.1 Team Member Visibility

This organization was utilizing a robust PPM solution, but locked its access to only the project planners within the organization. All inputs and outputs to the project plan went through one individual for each project creating a bottleneck for information sharing. The PMO audited some project mistakes and found resources working on tasks that were not on (nor near) the critical path, inefficient effort by each project planner to gather the status of work on the project, and sometimes even an inability to answer who was working on what project.

The organization needed a solution to support task management. They had two paths to choose between:

1. Implement a stand-alone solution with an interface to the PPM solution: The near-term tasks would be populated from the PPM solution into a task management solution, where all team members would be assigned and their progress tracked.
    - Pros: The organization felt that a dedicated task management solution would be well adopted by their team members
    - Cons: a burden on the project planners who had to utilize two tools in parallel to plan and track tasks
2. Utilize additional features of the PPM solution: Rather than open up the main project scheduling to all team members (an option this organization rejected from the beginning), utilize a simple access portal for individual team members to access their current work and update their status.
    - Pros: retain a single source of truth for active project related information, easing the project planner's day-to-day job
    - Cons: risk of additional data management within the existing PPM solution burdening the PMO, support staff, and application performance

Originally, the organization chose option 1 and their IT department built a homegrown solution. After a few years, they have reassessed and replaced option 1 with option 2.

10.2.2 Pipeline Complexity Requires Integrated Planning

A cutting-edge biopharma considered themselves a startup, though they had multiple assets in the market and a robust pipeline. Each function had historically worked in their own silo of project planning – each asset manager used a desktop project management solution, while each function utilized their own tools (from task management solutions to spreadsheets to back of napkins). Too often, a delay in one function's work was not visible to other functions causing cascading delays that could have otherwise been mitigated and avoided.

A project management solution with robust cross-functional planning capabilities to support the matrixed organization was required. Many times, individual functions resist transparency into their operating model ("don't look at my sausage making"), but a strong executive push can break down such silos. In this case, the organization's functional teams had actually been requesting an integrated solution, and both the PMO and functions were eager and able to set up the multi-level planning discussed in Section 10.3.1.2.2.

10.2.3 Operational Portfolio Management

A small but fast-growing biotech had its first program entering Phase 2 trials and additional assets entering the clinic and a few more promising molecules coming out of their research efforts. When the executive leadership team asked for updates on their active portfolio, the organization had to scramble to collect data. Given the fast-paced nature of their business, the expectation was set to be able to answer in real time the status of schedules, and eventually resource burn rates and bottlenecks.

The organization needed a project management solution that would scale with them – easy to use for basic tracking of Asset Development Plans (ADP) now and the ability to add additional integrated processes as they matured. The single output they sought from the beginning was a timeline view much like the second Figure 10.3 found in Section 10.6.3. They elected to use an off-the-shelf PPM solution and only utilize its project scheduling and reporting capabilities for an initial implementation but know that resource and cost management could be easily added later.

10.2.4 Finance Scrutinizes Project Forecasts

A "hot" organization with lofty expectations from Wall Street was under scrutiny for their cost projections. Finance lacked confidence in the forecasts provided for each Annual Operating Plan (AOP) cycle and often manually updated project budgets without assessing the impact. Meanwhile, the PMO and project management community were stressed as they spent considerable effort manually iterating with each function and collating spreadsheets to produce the project cost forecasts.

Finance was actually pushing for a non-project management solution that focuses on cost management use cases until they recognized the benefits of integrating their processes with the project schedule via a PPM solution. The ability to track assumptions, transparency into project metadata, and alignment of cost forecasting directly with the key work packages start dates and durations would actually provide more accurate estimates and more efficient planning turnaround times. The functional teams could enter their forecasts directly into the solution with a workflow tracking update status, freeing the project managers to focus on project planning and execution and not on data collection.

10.2.5 Resource Bottlenecks Causing Unnecessary Delays

A drug-device organization was growing fast, hiring resources left and right. But were they hiring the right resources, at the right times? They needed trusted project plans with resource forecasts beyond 1 year.

The organization implemented a simple PPM solution at first. Basic project management functionality was coupled with resource forecasting, but they were limited in their "what-if" scenario planning and could only plan at either the role level of their RBS or the named resource level (not both). They needed to (a) have dynamic plans that could help them assess how different development options impact the portfolio and (b) be able to plan at the role level in the long term (life of project) and replace with the named resource individual in the short term (current stage).

They switched to a more robust PPM solution, integrated it into their IT ecosystem, and have not looked back. After making the switch, the organization reports much more effective workforce planning: They have confidence with hiring the staff that will need to be trained and ready to deliver in one to two years' time. Meanwhile, the project managers are more efficient with the ability to model different "what-if" scenarios, as well as a well appreciated feature: utilizing the activity library to build out project plans (see Section 10.3.2.2).

10.2.6 Connected Strategy with Execution

A medical device organization had no problem with the delivery of each individual project but found their output did not match the decisions taken in portfolio review meetings. The balance between new products and sustaining support was out of sync. They liked to start new projects, but did not like to kill them, and they missed a few attempts to be first to market because they spread their resources too thin across all the active projects. The prioritization from leadership was not disseminated effectively to the organization. Gate review meetings did not use objective criteria. Resources were assigned based on who complained loudest, not to achieve corporate objectives.

The organization chose to increase transparency and integrate the business case of each program into their project management solution, including their priority score (they bucketed programs into three categories of priority). Right away, this enabled transparency to functional teams assigning resources to ensure that the highest priority programs were staffed appropriately. Additionally, gate review meetings followed a more standard protocol of comparable KPIs and the evolution of the business case could be evaluated over time for trend analysis that aided in making informed decisions about whether the project should continue to the next stage or not.

10.3 PROJECT PLAN CONSIDERATIONS UNIQUE TO LIFE SCIENCES

10.3.1 "What Is a Project?"

Complex drug development projects require a thoughtful approach to how work is organized and structured. The first question to even consider is: "what triggers a project to be a project?" Formalizing a project management solution forces organizations to consider this question they otherwise may not have addressed before. For many organizations, this question may seem silly – maybe Finance dictates a project code at a certain process step (e.g., Request-for-Development milestone achieved), or a project manager is assigned at a certain point in the research or development process. If such examples are not clear in a project team's process, creating a template will likely force the issue. I recommend ensuring alignment between project teams, functional teams, and especially Finance so that all teams are planning in concert together.

10.3.1.1 Project Structure

More fundamental than the starting gate of a template is the structure of the drug development planning paradigm, which requires answering the question "**what is a project?**". Utilizing a project management software system will force an organization to define separate project plans with some intention, and this structure must align with an organization's roles and responsibilities. Tip: consider a software tool with flexibility, because even in the most mature organizations, these roles and responsibilities change over time and your software solution should support an evolving business process landscape.

Consider these most common project structures: one project per asset development plan, one project per indication, one project per asset and one project per clinical development plan, and one set of interconnected projects per indication.

10.3.1.1.1 One Project Per Asset Development Plan

Organizations structure their definition of a project to align with the asset (or molecule) when either their focused therapeutic areas rarely have additional indications (e.g., certain cardiovascular programs) or when their aim of a PPM solution is for very high-level planning.

10.3.1.1.2 One Project Per Indication

The most common project structure implemented by biopharma organizations is one drug development project per indication. With this definition of a project, a subset of activities – and thus resources and cost – must be planned and considered: the work done one time to support any indication within the asset (for instance, early investment activities such as non-safety and non-clinical work or manufacturing considerations such as tech transfer work). This is commonly addressed by tracking this work on the "lead" indication – a rather simple approach that often aligns with Finance's accounting. This approach requires a template that may be different for lead indication than follow-on indications (which may already be the case anyway). In the rare instance that the lead indication project is either terminated or delayed, those shared activities, resources, and cost need to be manually shifted to the new lead indication.

10.3.1.1.3 One Project Per Asset and One Project Per Clinical Development Plan

When an organization has a project control processes at the asset level, distinct from those control processes at the indication level, then having two distinct projects for each level makes sense. This structure is evident when organizations assign separate project managers to each project type.

This project approach solves the challenge of one project per indication with regards to the subset of activities which is done one time to support all indications within the asset: In this structure, it is clearly planned within the ADP.

10.3.1.1.4 One Set of Interconnected Projects Per Indication

When functions lead the planning efforts, without centralized project or program management, an organization's definition of a project is each functions' work. Only collectively do each of these project plans drive asset, indication, or even clinical development plans.

Even with interproject links aligning the dependencies between functional plans, this definition of a project often results in a lack of transparency within an organization. Care must be taken to ensure that the same activities are not being tracked across different project plans (with the potential for having different constructs and different reporting of the "truth").

I see this structure as a starting point with fast-growing organizations who have at most a handful of assets and indications. Integrating these functional projects into a coherent project management framework often comes hand in hand with developing a PMO and investing in PPM processes and software solutions.

On the other hand, very mature organizations marry together centralized planning and functional planning, as discussed in the next section.

10.3.1.2 Functional Plans

Within the project structure of having a centralized asset and/or indication project plan, a question remains where to plan and track the functional work.

The basis of this question comes from the distinction between planning (centralized) and execution (decentralized). Responsibility of work package delivery resides outside of the centralized project manager and instead with functional teams. For many organizations, these functional teams are the "doers", but are not accountable for the planning of the work; for other organizations, some or all of the functional teams have accountable planners for their own work, often formally in a project manager role for the functional work. In either case, the work done by each function is highly dependent on other functions, with the management of these dependencies a key to success in managing full program scope, budget, and timeline.

Should all of the functional work be contained within the asset and/or indication project with one integrated project plan, should the functional work reside in coordinated separate project plans (a.k.a., cross-functional planning), or should the functional work be left to the functions to plan in their own dissociated manner?

10.3.1.2.1 One Integrated Project Plan

Some project managers create a single project plan with all activities contained in a single project with one work breakdown structure. With drug development, these project plans often balloon to many thousand lines of WBS.

Good for:

- Organizations with strong centralized project management function, and less planning accountability throughout the rest of the organization
- Critical path analysis – all relevant activities are available at your finger tips
- Very clear single source of truth

Challenges:

- The number of activities is often so large as to create practical challenges including
    - Critical path may meander through tasks that are not actually core to execution but appear to be thanks to complex chain of links or constraints.
    - Changes to one work package (e.g., extending a duration) can have unintended downstream consequences to the project plan due to complex chain of links.
    - What-if modeling becomes very cumbersome with many activities to manage.
- Ownership of activity, deliverable, cost, and/or resource does not reside with the project manager who has access rights.

Mitigations:

- Some organizations have created a job function for managing the project plans, distinct from the traditional project manager. I have seen this created job function literally be called "project planner", whose job duties are to be the only person touching a complex project plan within a software tool while the traditional project manager handles the rest of the typical project manager duties, such as defining strategy, coordinating activities, managing risks, and managing the team.
- Organizations that track a large number of activities in their project plans often regularly evaluate their templates for opportunities to simplify, asking if each individual activity is really required for active management.
- Some organizations will separate activities that are only used for resource management from those that the project manager uses for tracking critical path, to decrease the scope of focus of the project manager on their project plan.

10.3.1.2.2 Cross-Functional Planning

Some project managers create a high-level integrated project plan (IPP), with details broken out into separate functional project plans, with clear transparency and dependency between the IPP and the functional projects.

The classic example of cross-functional planning is with clinical studies, where organizations are most likely to have dedicated functional project managers. Within the IPP, the clinical study is planned at a high level, with maybe 10–20 milestones listed and tracked. Separately, a clinical study manager plans the study at a detailed level, often with hundreds of activities, including the same 10–20 milestones as the IPP. Thus, in two separate project plans, the same 10–20 milestones are replicated. The replicated milestones are indeed duplicates from a planning perspective and are kept synchronized to ensure alignment between the IPP project manager and the clinical study manager (Figure 10.1).

In addition to allowing for scheduling to be democratized to the teams doing the work, resource estimates and/or cost planning may also be planned and executed by

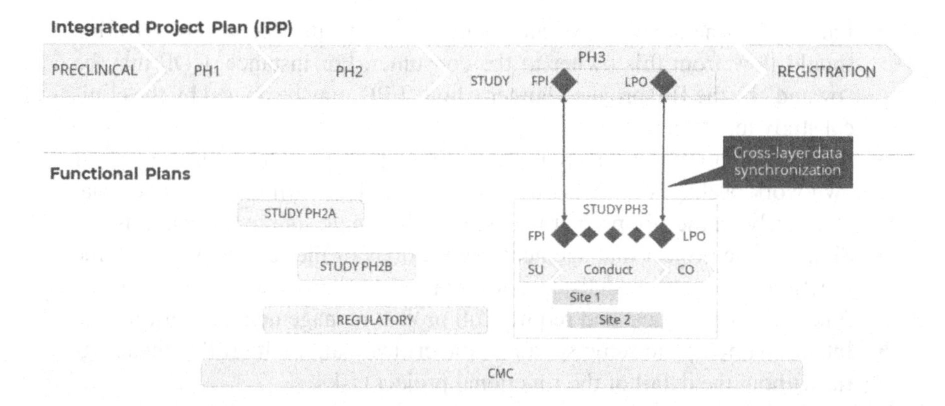

FIGURE 10.1 Visual representation of the crosslinking of functional plans into the integrated project plan (IPP).

the functional teams, integrated to their schedules. These resource and cost plans can then be consolidated within the IPP for capacity planning and budgetary planning purposes.

Good for:

- Organizations with strong project management expertise within the functions
- Simplifying the planning and tracking of the IPP while not losing the benefits of detailed work package management
- Bringing transparency throughout the organization by breaking down siloed planning which may have been occurring through disparate tools
- Empowering functions to be accountable for their deliverable planning
- Organization time/energy is spent planning in detail "just in time" – the IPP planning is done far in advance at the higher level, and the functional plan is created for execution.

Challenges:

- The milestones that are duplicated into two separate project plans risk being out of sync
- Project management expertise is less likely to be found within the responsible resources within the functions (a real-world example: A good statistician is promoted to lead the biostats function and is responsible for the functions planning, but has no training on project management practices).
- Detailed critical path analysis flows through separate project plans.

Mitigations:

- Synchronize the replicated milestones via a purpose-built software solution, rather than relying on home-grown solutions.

- Each milestone should have an "owner" of its truth. The synchronization should flow from this owner to the consumer. For instance, CDP may be "owned" by the IPP project planner while "FPI" may be owned by the clinical study manager.
- A hybrid solution is often appropriate: Some functional teams manage their own work packages in their own separated projects, while other functional teams rely on the IPP project manager as the single source for planning.
- Simplify the project management expectations of the functions – for some functions, perhaps a task list is more appropriate than a Gantt, whereas other functions may request and require full project management capabilities.
- Intentional template setup such that the critical path at the IPP is meaningful without the detail of the functional project tasks.

10.3.1.2.3 Dissociated Functional Plans

Left to their own devices, some functional teams will utilize their own project management solution, whether via a formal software supporting the functions' planners or each responsible functional team member tracking in their own Excel or PowerPoint.

Good for:

- Functional teams track the specific resource to each work package and need to manage the timelines in order to do so effectively
- Functional teams are culturally used to working in a silo (I have heard, "I don't want others in the organization to see our sausage making")
- Functional teams that are not given the opportunity to track the level of detail needed for their own management and therefore feel forced to use their own tool

Challenges:

- Lack of transparency across the organization and separate sources of the truth – the IPP project manager is tracking and reporting to one date while the function may be planning to another
- Additional software solutions supported by the organization (increasing IT effort, contract management, vendor management, training, etc.) when a function purchases their own solution
- Decreased integration between functional teams for coordination

Mitigations:

- Functional representation within each project team with dedicated time in team meetings to review individual functional timelines
- Executive management push to break down silos and encourage cross-functional alignment

10.3.2 Use Templates

A value of any project management tool is to standardize the starting point of the project plan, that is, with templates. The project manager gives up the blank canvas of doing whatever the he/she desires, but in return gets:

- time savings during the project creation process
- consistency from one project to the next
- latest and greatest best practices from process improvements incorporated into template updates
- avoidance of bad habits perpetuated from doing save-as from previous projects

10.3.2.1 Success Factors

Here are various success factors I have seen when building out their templates:

- Fully flesh out the answer to "what is a project?" before embarking on template definition
- Creation of templates should not be left solely to the project managers. As will be discussed below, cross-organizational considerations should be factored in to have a successful set of templates, including functional libraries, alignment with resource manager teams, alignment with Finance, and organizational buy-in for the level of detail. This is why organizations have a PMO: to address such cross-divisional process alignment.
- Avoid a proliferation of templates. It is easy to succumb to the specificity of various project types (lead indication vs follow-on, each therapeutic area, structure of clinical development plan, etc.). In the long term, the pain of updating each template will outweigh the benefit of the project managers' project creation, and as templates are just an estimate of a project plan anyway, being too precise risks missing the mark. Indeed, if templates are clearly not exact, the project manager will be forced to provide their subject matter expertise and update their project plan as opposed to relying on the project to be exactly aligned with the template (or being afraid to update the template). I have seen very large pharma organizations be efficient with two templates: one for large molecule drug development and one for small molecule drug development.
- Build activity libraries. Even if the PMO builds a whole detailed project plan for template creation, the concept of activity libraries is a huge benefit for project managers. Consider the case of adding a new study: The project manager could copy/paste an existing study from their plan, which may not be the appropriate study approach (e.g., copying a Phase 1 study into Phase 3), may have data attributes which should not be copied (example: study protocol number), and may not utilize the last process improvements. Instead, if the project manager could go to a list of possible studies, and choose the one that is applicable for them, in a few short clicks, they could build out the accurate study types. Advanced PPM tools even ensure that inserted activity libraries are linked correctly to the target existing activities.

- Define the appropriate level of detail. I have heard multiple times that even across a 10-year cross-functional program plan, only about 30 milestones really drive the critical path. And yet, such plans can have literally thousands of activities. I encourage the level of detail to not be a foregone conclusion, but to be an intentional decision.

10.3.2.2 Template Level of Detail

Intuitively, templates contain a whole detailed project plan, start to finish with the most common WBS contained within. Should they? An alternative template structure would be to contain only a skeleton set of activities, such as the standard set of phase and gates, or only the main work packages within each phase. Project managers could then choose from a "library" of activities to insert into the project as a sort of building blocks. These activity libraries are predefined with WBS, durations, and link dependencies between them, just like a project manager would expect from a template.

This approach has the following benefits:

- Flexibility for the project manager in choosing the building blocks from the activity libraries to easily tailor the project plan for the project's specificities
- For long project plans, a recognition that the ways of working many years in the future may change
- Enable easier what-if scenario planning
- Encouragement to do long-range planning at a higher level, as opposed to trying to detail out resource, cost, and durations on tasks very far in the future
- Ability to add detailed WBS "just in time" – typically one phase in the future, which ensures the latest updates in process when importing the activities from the activity library

10.3.2.3 Template Resistance

I have heard a resistance to being "forced" to use templates from a project manager that their projects are unique and don't lend themselves to a template – especially with new-to-the-world medical devices or clinical development plans that are taking an innovative route. I encourage these resistors to give templates a try and feel free to modify the project as much as desired after initial creation: add and delete WBS, shrink and increase durations, adjust dependencies, etc. There is very often structure contained within a template that the project manager will realize is helpful. As a last resort, PMOs could provide a "blank" template – a template without any WBS but still aligns metadata (for instance, a corporate calendar, system access rights, naming conventions, etc.).

10.4 RESOURCE MANAGEMENT INTEGRATION WITHIN PROJECT MANAGEMENT TOOLS

As acknowledged in Section 1.2.7, a robust resource management tool is often required to do the job efficiently. The most oft-used resource management tool is not robust: a spreadsheet. The most common, and effective, tool for resource management is actually the project manager's tool: integrating scheduling directly with resource planning. The benefits of integration include the following:

- Elimination of silos between schedule and resourcing, which are highly interdependent data
- Visibility to the project manager of the resources required to execute his/her project plan
- Inherent communication of when work needs to be accomplished to functional teams who are accountable for the resources' execution
- Ability to efficiently perform what-if scenarios and communicate the resourcing impact to functional teams to enable improved organizational decision-making between possible scenarios
- Earlier risk identification of resource bottlenecks
- More likelihood to have the right resources working on the right tasks at the right time – thereby improving the project manager's ability to ensure project delivery success

Even at organizations where project managers do not mingle resource management within their project management processes, it is best practices for resource management to be actively integrated into the project plan and project toolset.

Successful integration of resource management and project management practices requires answering a number of process questions, related to the WBS, RBS, definition of FTE, resource distribution, project calendar, and who plans the resources.

10.4.1 What Level of WBS to Assign Resources?

This question addresses where within the project plan resource assignments should be created. In asking this question to stakeholders at organizations who are revamping their template or implementing a new solution, I generally hear conflicting answers, but each stakeholder is very confident in their answer! The main answers are at the project level (i.e., resources are assigned irrespective of any activity), at an intermediary work package level (e.g., at the study or equivalent level within an ADP), or at the detailed task level (i.e., the lowest level within the WBS). The following considerations should be addressed:

- Project Level
    - Pros
        - Easy to enter – typically resources are assigned in monthly, quarterly, or annual buckets
        - Resource plan is independent from the schedule. This independence may be appropriate when the resource management processes are completely separate from the project manager's scope.
    - Cons
        - Changes in schedule have no impact on the resources (for instance, if a study is delayed, the resource plan has to be independently manually updated)

    - Difficult to understand the logic behind a resource plan (for instance, medical writers are required leading up to submissions, but one has to scour the schedule to see the submission milestone to understand this connection)
    - Inability for a project manager to easily run what-if scenarios
  - Considerations
    - Assigning resources at the project level effectively divorces the resource management from the schedule management
    - Could be a first step in getting an organization to recognize benefits of integration, without disrupting existing processes too much
    - I see project level assignments more often within medical device than drug development organizations
- Intermediary work package level
  - Pros
    - Goldilocks approach to provide an integrated solution between resource management and scheduling, without requiring non-value add effort on the project manager and PMO
    - Often aligned with the level of discourse that the functional teams are thinking of resource management
  - Cons
    - Risk of inconsistent rules with resource assignments happening all over the project plan
    - Separation of reality: The resources are actually doing real-world work on the tasks the project manager is planning
  - Considerations
    - For some organizations, hammock tasks are created between two milestones to house resource assignment, which serve the purpose of tying the resource plan to the schedule, without impacting the project manager's dependencies
    - I recommend aiming for consistency with the level of detail used for resource management. For instance, try to avoid clinical trial managers planned at the study level, while medical writers are planned as subtasks of the study.
- Detailed task level
  - Pros
    - Adheres to PMI best practices[1]
    - Precise impact analysis on resource plans when making schedule changes, either in what-if scenarios or directly on the project plan
    - Distribution of resource assignment most accurately tracks reality (see section 10.4.4 below)
  - Cons
    - Onerous data management on the part of the project manager maintaining the resource assignment within the project management tool

[1] https://www.pmi.org/learning/library/schedule-101-basic-best-practices-6701.

  - Often, functional teams do not think of their work in the independent tasks that a project manager tracks but rather broader work packages.
- Considerations
  - This method is successful when the template and expectations for the project manager is to be less granular in their schedule management. For instance, a clinical study with 20 tasks can have effective task-level resource management in a way that a clinical study with hundreds of tasks cannot.

10.4.2 What Level of RBS to Assign Resources?

To some people, "resource management" means the assignment of individuals, while to others it means the role or skillset that will do the work, while to others still, it is for financial purposes and therefore refers to assigning the cost center from which the work will be executed. This potential confusion must be clear from the beginning of discussing an integrated PPM solution. I have observed very strong feelings about this topic among various stakeholders, as well as wrong assumptions about what peer stakeholders were thinking when they used the term "resource management".

The most common approach within drug development is to plan resources at the role or skill level and assume that the individuals are fungible. Finance's needs are placated by ensuring the RBS aligns with their method of tracking cost centers.

For some organizations, individual named resource assignment is done for a predefined subset of roles, for instance, clinical site monitors. However, I have observed named resource assignment across all R&D roles; extra process controls and support must be in place to be successful because of the added data management across the organization. Full named resource management is more common for medical device organizations and for CDMOs.

10.4.3 What Is the Definition of an FTE?

When a project manager says "I need 1 FTE to complete this task for a year", does that mean one person (or two half-time people), or does that mean more than one person is required recognizing that each individual has vacation, training, functional project initiatives, and other non-project working time? Said another way, does 1 FTE equal 2080 hours per year (40 hours per week times 52 weeks) or more like 1664 hours (which represents 80% working time)?

I have observed organizations take more than a year to align and answer this question internally! Alignment matters, because if a project manager reports on the number of FTE planned, the functions need to have the right staffing and finance needs to know how to monetize the work – if each constituency is 20% off from the other, the value of integrated planning diminishes.

I think there is no "right" answer. Take the time to harmonize within your organization, be sure to explain the definition in your trainings, and stick with it.

10.4.4 How to Distribute Resources across an Assignment

A resource assignment is flat by default. For instance, a six-month task requiring one FTE will require one FTE each month. Is a flat distribution always appropriate? What if most of the work is done in the second and third months? The classic example is the effort involved on a clinical study: Startup requires quite a bit of work, the major resource load is during patient enrollment, and then, the resource needs decrease in closeout and submission.

One option is to manually estimate the resource effort month over month; while precise, this requires quite a bit of organizational maintenance and loses the connection between the resource assignment and the task. Most biopharma organizations instead rely on a compromise of detail, where the most meaningful milestones drive resource-assignment work packages. These work packages may or may not be the same tasks that the project manager is already tracking; in the case that they are not, new activities must be created within the project plan.

The graphic below shows a typical setup for clinical studies: Just four work packages are used for assigning clinical operations resources within the study. The result is a system that is easy to maintain, while still capturing the major drivers of distribution (Figure 10.2).

10.4.5 What Calendar to Use?

With stand-alone task or project management solutions, a project manager can decide to schedule over 365 days, 260 days (5 days per week over 52 weeks), or exclude company holidays as working days. In a global environment, the holidays are harder to maintain (e.g., is my team impacted by Chinese New Year). With an enterprise project management solution, the calendar options may be fixed for project managers.

Even though I have observed hard-core PMI schedulers demand precise calendars with every holiday tracked, I recommend keeping planning calendars to the two most basic: a 260-workweek calendar and a 365-day calendar for when activities are

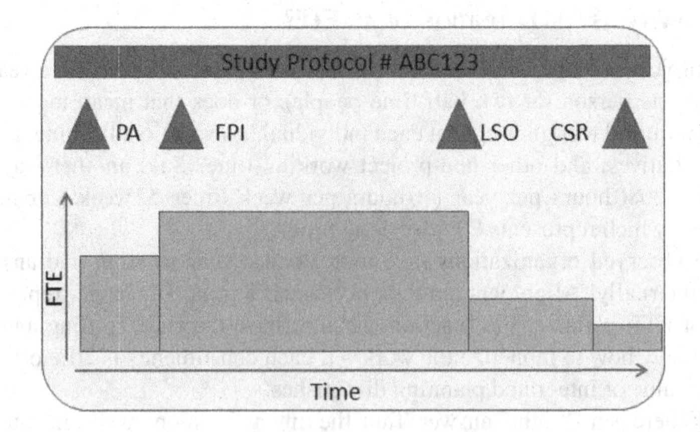

FIGURE 10.2 Visual representation of the typical allocation of Clinical Operations resources (FTEs) across the clinical trial lifecycle, using key milestones as inflection points.

known to be executed on the weekends (for instance, often related to manufacturing activities). This is more intuitive for less experienced project managers, is easier to harmonize across the organization, and reduces the potential for confusion on how a system is setting up dates.

10.5 THE BROADER ECOSYSTEM: IT DATA FLOW

The project manager's effectiveness is enhanced when data flows transparently, silos are broken down, and time is not wasted trying to gather or provide information. As such, it is in the project manager's interest to have their project management tool automatically pass data between various other systems within their organization. The most common interfaces within drug or device development are with the clinical trial management system (CTMS), enterprise resource planning (ERP) systems, reporting solutions, and to share data with clinical research organizations (CROs).

10.5.1 CTMS Integrations

CTMS tools exist to capture execution details of clinical studies. Examples include IMPACT, Veeva CTMS among many others. They capture information about trial participants, enrollment details, and, critical to the project manager, study milestones. These systems are heavily regulated by the health authorities (for instance, to have every data change auditable and archived), detail-oriented, focused only on study details, and do not adhere to PMI best practices. Furthermore, CTMS solutions do not afford the opportunity to consider what-if scenarios, do resource or cost forecasting, and typically focus only on known studies in the near-term rather than long-term portfolio planning of studies. For these reasons, I do not recommend clinical project managers relying on their CTMS as their project management solution.

A study needs to be considered for portfolio planning and even for asset planning within the project management solution prior to being tracked within the CTMS solution. Therefore, metadata about the study is tracked from a planning perspective and can be provided to the CTMS solution when needed, reducing the need for double entry and possible misalignment between tools.

Even though a CTMS tool is not suitable for proper project management, it is the source of truth for the actual milestone dates and possible other key study information. Both a clinical project manager (if one exists) and the development project manager must utilize this data in any of their planning and reporting. Therefore, these data can be provided to the project management solution when created and when updated, again reducing the need for double entry and possible misalignment between tools.

For many organizations, a bi-directional software interface is created between their project management solution and their CTMS with the following data considerations:

- Study identifier (name or code)
    - Placeholder created in project management solution
    - Actual identifier created in CTMS

    - Placeholder in project management solution replaced by actual identifier automatically
- Milestone dates
    - Planned originally in project management solution
    - Planned dates automatically sent to CTMS when study created within CTMS
    - Actual dates tracked within CTMS and automatically updated within project management solution
- Fixed study metadata, such as therapeutic area, phase, etc.
    - Entered and owned within project management solution
    - Information sent to CTMS when study created within CTMS
    - Data locked (cannot be modified) in CTMS
- Variable study data, such as number of subjects, number sites, etc.
    - Planned originally in project management solution
    - Data automatically sent to CTMS when study created within CTMS
    - Actual data tracked within CTMS and automatically updated within project management solution

10.5.2 ERP Systems

PPM solutions are to the PMO what ERP solutions are to Finance. In other words, Finance organizations rely on ERP solutions for their core business function; I have never encountered an organization mature enough to invest in a PPM solution which did not already have an ERP solution in place (though I have encountered many who revamp their ERP solutions, an implementation endeavor that is often years long). ERPs are more often than not software from SAP, Oracle, or Microsoft. Typically, an organization's budget is recorded within their ERP. That being said, the budget is best *enabled* through forecasting within a PPM solution.

ERPs should remain the system of record for the project budget, but this system of record should not require double entry: It can be effectively enabled via an automatic data flow interface from the plan in the PPM tool to the ERP solution. The process question must be answered as to when the planned budget in the PPM solution should be sent to the ERP – for some organizations, this is calendar-based (i.e., defined ahead of time to be the current plan on a specific day), while for other organizations, it is based on a trigger within the PPM solution such as creating a particular baseline or passing a gate review with a "go" decision.

On the other hand, the actual costs are booked within the ERP; this information is vital to the project manager in order to make short term decisions (e.g., dealing with cost overruns), to improve long-term forecasting processes by the PMO, and for proper reporting on the project. Actual costs are often therefore sent from the ERP to the PPM solution on a monthly cadence.

One gotcha for the project manager to be aware of: SAP's cost structure adheres to a WBS that is often structured differently than project teams, requiring creative mapping for the data to automatically flow and often resulting in a coarser level of granularity within the ERP than was used to do the planning in the PMO solution. For instance, I have observed many organizations whose SAP tracking of project costs is at the trial phase level for Clinical Operations, which inhibits analysis at the study level, let alone any key work packages within the study.

10.5.3 Reporting – Output and Access

As seen in Section 9.4.3, the project manager's ability to communicate to stakeholders is enhanced with a proper reporting tool. Larger organizations often set up a data lake or data warehouse as repository to run their reporting software off of. The project data must feed into this repository for corporate reporting purposes, so data interfaces are created to populate the repository on regular cadences (most often, daily, though sometimes more frequently).

With this reporting ecosystem where enterprise dashboards are setup and maintained in a different tool than the one used for project management, I recommend the project management community retain control over some key reporting within their own system. For instance, with project status updates, or graphical timelines or roadmaps, the project manager benefits by being nimble and able to have updates from the project plan immediately reflected in the reports rather than waiting for data to sync across systems. Furthermore, when the PMO or even an individual project manager wishes to update a dashboard or layout, the technical change required may take weeks, months, or even quarters to get pushed through the IT processes.

10.5.4 CROs

When a project manager's program includes work contracted to a CRO (or a CMO, for that matter), the project manager loses some control on project status, reporting capabilities, and cost controls. While regular meetings with the CRO can provide unofficial information to the project manager, the CRO usually communicates formally at the end of each month with their key data (such as patients enrolled, passthrough costs accrued, contracted milestones achieved, etc.). Furthermore, these data are often communicated through procurement processes to Finance first, before being transparent to the project manager.

Integrating the project management solution directly to the CRO's operation team opens the door for increased transparency to the project manager and project control. Some project management solutions such as Planisware and Smartsheet enable project managers to send data requests such as status updates, percent complete, and units accrued from within their software to external parties. Such data integration systems are controlled so the CRO can only see the data precisely relevant to them, and the project manager can validate the data via workflow before accepting it into their plan.

Even less transparent to a project manager is when a change in contract is needed, for instance, to activate additional sites. Not only does the cost and timing impacts often come late to the project manager, the project manager's organization is often not in a position to validate a change in costs. With an integrated system, the requested change by the CRO can be communicated to the project manager who then can run what-if scenario models to understand and mitigate the impact of the change to their project as well as model the impact to costs and validate that a change proposal by the CRO is acceptable.

10.6 TIPS FOR SUCCESSFUL CHANGE MANAGEMENT

People, processes, and tools: No change is successful without all three coming together with intention. This chapter has focused on the tools and specific process considerations when implementing a project management solution – the people are equally important. This means that the community of project managers who will be the tool primary end-users, but also the data consumers, the impacted team members, the functional areas, and the sponsors.

Often, the term "change management" is addressed simply with a plan for initial training. As this section will show, it is much more than a single training session.

10.6.1 Executive Stakeholders

The number one key to success for an impending change is to ensure that executive stakeholders are bought in. Early and frequent communication of the business case, status, and value being provided will make it easier to secure budget for a tool, ensuring that priority is given to resources to work on the implementation, navigate cross-department misalignments, and allocate support of the solution in the long term.

Once implemented, the executive stakeholders have an even more critical role to play in the success of any changed process and tool: utilizing the data to make their decisions. This sounds obvious or simple; it is not. Executives should utilize the reporting provided, request data to come from the implemented solution, follow any process workflows, and support roles and responsibilities that have been set up.

Especially in the early stages of change being implemented, data within a project management solution could be incomplete, inaccurate, or not current. In such cases, what should an executive decision maker do? They could request or expect the data to be manually updated in a spreadsheet or a slide, but doing so undermines the implementation. The single most powerful way for data to be correct is for an executive leader to call out incorrect data in a meeting: I have heard of a head of R&D cancelling an executive leadership team meeting because a project manager's schedule did not reflect a recent decision. I am sure that project manager updated their project plan right away next time!

10.6.2 "What's in It for Me?"

For each stakeholder, ensure that there is an answer to the question of how they benefit by the change. Seek out a list of these benefits. Tell them early and often.

For the project manager, it may not be obvious what they gain if they have to give up their favorite stand-alone project management tool they have used for potentially decades when implementing a new project management solution with imbedded PPM processes. Here are some options to consider:

- Find unique functionality the new tool offers (e.g., unlimited baselines)
- Reports or dashboards built for the project manager, saving them time to recreate manually

- Self-service to the plan from stakeholders means they no longer need to bug the project manager at all hours of the day or night for updates
- Highlight the interconnected nature
  - Good resource management alleviates bottlenecks – and headaches – for the project manager
  - Cross-functional planning provides early warning to the project manager of deliverable misalignment
  - Integrated time tracking illustrates when project team members actually start doing the work the project manager expected of them

10.6.3 Begin with the End (Report) in Mind

Although the primary purpose of a project management solution is to support project execution, an increasingly important purpose is to report the status of projects to senior leadership. With that in mind, it is helpful to consider the types of reports that your PMO will generate and then build the solution to accommodate those reports (Figure 10.3).

10.6.4 Set an Implementation Up for Success

It should be clear by now that the effort of implementing a project management solution should not be underestimated. The best mechanism to be successful in change management is to have a successful implementation. Here are a number of best practices:

- Do not boil the ocean
  - Consider an minimum viable project (MVP) where not all functionality is released from the beginning. A possible progression may be ADPs focused on scheduling, then functional plans, then resource

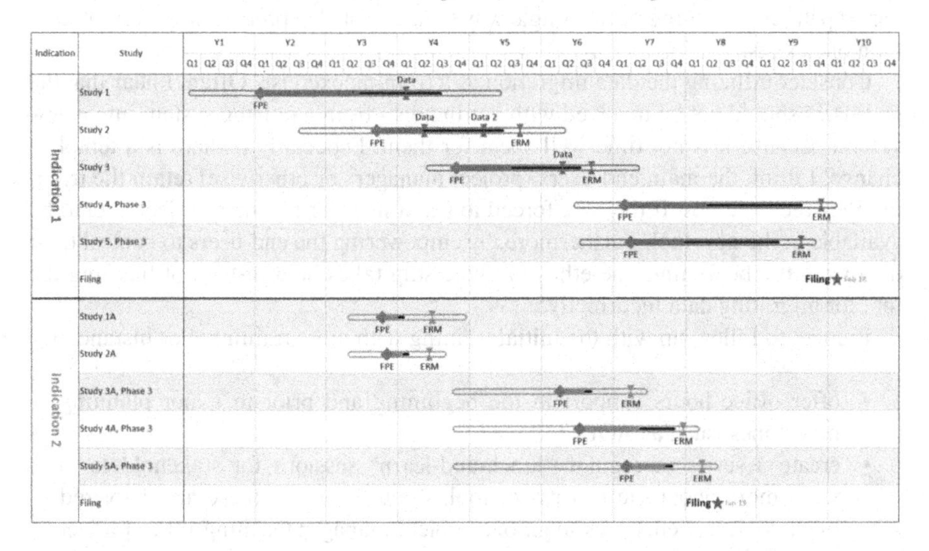

FIGURE 10.3 Example roadmap that can be used to show the entire development plan for an asset.

management, and then closed loop planning with actuals from time tracking
    - Pilot with some groups before expanding the user community (for instance, implement for one therapeutic area before the others, or for Clinical Operations before the rest of the functions)
- Work with experts in the field, either directly with the solution provider's team or via an external consultancy. It is expensive, but less costly than a failed implementation. I often hear from the customer team, "but we can't make the right decision because we don't know what we don't know" – exactly, so hire the folks that do know.
- Select your core team carefully: They should be empowered, representative, and trusted within the organization. And also have availability to focus their energies toward the implementation effort. Many times, no one fits the bill; this is where executive sponsorship is important, because the right people's priorities can be adjusted to make them available.
- Don't neglect post-launch support
- Know your processes
    - Have a realistic assessment of your organization's process capability. Consider a maturity assessment to begin.
    - Do not underestimate the organizational alignment needed in defining the data model and data dictionary

10.6.5 Training Never Ends

A good training plan is essential. It should not focus on the tool, but the process change; where the tool is highlighted as the enabler for the process. That being said, of course the tool needs to be clearly described. For initial trainings, focus on the happy path, highlighting only a single way to accomplish a process step even if more than one exists.

Consider utilizing the data migration as a training exercise. Often, I hear that the end-users should not be involved with getting data from a retiring system into a new system because it is not time well spent for them. Especially if there is a template change, I think the main end-users (project managers or otherwise) retain the training they received best if they are forced to use it in order for their project data to be available in the new tool. Furthermore, by empowering the end-users to control their data from the beginning, the ethos of ownership takes hold instead of blaming the tool for migrating data incorrectly.

Prepare to follow-up with the initial training with more training. For instance:

- offer office hours support in the beginning and prior to major planning milestones (such as AOP)
- create a series of regular "lunch-and-learn" sessions for stakeholders to learn more about the process or tool. Bonus points if these are delivered by end-users themselves (e.g., one project manager teaching other project managers a cool new feature)

And do not overlook new users onboarding over time, either new hires or those changing roles internally. The training they receive should be no less rigorous than at initial deployment and often must contain additional process training. For instance, a new clinical trial manager who has to update the tool may first need to learn the basics of project management.

10.6.6 Tying It All Together

Change management starts at the beginning and never ends. Organizational pain points must be articulatable to seed rationale behind making a change. Program status can be included in all-hands meetings and added to project team meetings. Track the benefits and communicate them over time; especially beneficial is finding stories to share about the changed process operating in real life. Continue to engage the executive stakeholders.

With intention in the communication, the people adopting the changes will be prepared for the process and tool changes. Even abiding by all best practices, change is not easy. Ultimately, as long as the corporate goals are kept in mind – getting therapies to patients faster, producing life-improving medical devices, or whatever it may be – the effort will be worthwhile.

10.7 SUMMARY

Project managers need a good project management tool. When assessing tools, it is important to consider how project information will be made available to stakeholders, how cross-functional workstreams will be integrated, how the solution will scale as the business grows, how cost and resource management will be included, and how projects align to corporate strategy and prioritization. Owing to the unique nature of drug development projects, there are additional considerations to sort out before implementing a project management solution in biopharma. The company should define the work packages that will be tracked and create project plans that address those work packages. These work packages can also be used to manage human and financial resources. The project management solution should be able to pass and receive information from other line-of-business applications such as the CTMS, ERP, financial planning, reporting tools, and external sources such as vendors.

Rolling out a new project management solution requires dedicated change management tactics. It is essential to obtain executive buy-in to the project before you begin and to have a clear value proposition ("what's in it for me?"). It is wise to think ahead about what you want to get out of your project management solution, often in the form of the reports that will be used to communication project status to stakeholders. You may want to start small and build on quick wins, such as piloting a few project plans in one function before rolling out to others. Training is also essential, and training needs to be done continuously with those interacting with the system. All of these tactics will help to make the implementation of a new project management solution successful.

11 Creating an Asset Development Plan

Joseph P. Stalder
Groundswell Pharma Consulting

CONTENTS

11.1 INTRODUCTION

Bringing a new drug to market is an expensive, complex, and high-risk endeavor. The best chance a project team has to make the process smooth and successful is to have a comprehensive plan that integrates the strategic considerations from key functions across all stages of development. The resulting document, called an Asset Development Plan (ADP), becomes the project team's guide for how to maximize the value of an asset.

This chapter discusses the purpose and contents of an ADP. In the next chapter, we will explore the process for creating a Clinical Development Plan (CDP), which provides a plan for how to achieve regulatory approval for one indication within the ADP. Note that for some drug candidates and for some therapeutic areas, there may be only one indication for an asset, in which case the CDP is contained within the ADP. In the example below, we will assume the asset has potential in several indications, as it happens in oncology or inflammation.

DOI: 10.1201/9781003226857-13

11.2 WHAT IS AN ASSET DEVELOPMENT PLAN AND WHY SHOULD A PROJECT MANAGER CARE?

As the guide for maximizing the value of an asset, the ADP focuses on areas to generate revenue and to reduce costs. Key revenue generators include the market opportunities or indications that will be pursued, commercial strategies to secure and grow the market potential, and legal protections for the asset's intellectual property. Key cost reductions include reducing the estimated cost of goods (COGs) via manufacturing process improvements.

An ADP is valuable for biopharma companies of all sizes and programs of all stages. By laying out a complete plan, project teams can

- Align with strategic objectives set forth by senior management
- Define phase gate decision points for development of the asset
- Outline major development milestones and success probabilities that can be used for project portfolio prioritization and management
- Align functional work streams within the program and establish efficiencies in the program that can reduce timelines and development costs
- Identify risks that could delay or terminate the program
- Improve the odds of successful regulatory approvals.

The development project manager is the project team member best suited to drive the creation and maintenance of the ADP. The development project manager has cross-functional and cross-indication visibility of the asset's work packages. The development project manager has the most to gain from having an ADP. The integrated, consolidated plan helps the project manager in that it becomes the "single source of truth" for cross-functional planning. It contains the assumptions that are used to align the project team to a common objective and the agreed objectives with senior management.

11.3 WHAT IS IN AN ASSET DEVELOPMENT PLAN?

An ADP has the following sections:

- Product vision statement
- Summary of commercial forecasts across all indications
- CDP
- Regulatory plan
- Nonclinical development plan
- Chemistry, Manufacturing, and Controls (CMC) plan
- Medical affairs plan
- Commercialization plan
- Product protection plan

In the section below, we will walk through the key sections that the development project manager should know in order to drive the successful creation and maintenance of the ADP. Your organization's business processes will define when updates to the ADP are required. For example, some companies have an annual or twice annually portfolio review process; some companies require an update to the ADP after every governance interaction; some require updates only when strategic changes to the plan are made.

11.3.1 What Is a Clinical Development Plan?

In the context of the ADP, the CDP lists all the indications that will be pursued and the development path to obtaining market approval for each indication. Chapter 12 describes in detail all the content of a full CDP. The CDP section of the ADP, then, contains a subset of the information from the CDP. I like to use a high-level roadmap visualization to start this section of the ADP, such as below (Figure 11.1):

Next, I summarize the development plan by budget and development phase using three categories:

- **Committed pivotal**: All activities to support the indication are in the budget and fully resourced.
- **Committed exploratory**: All activities to support the exploration are in the budget and fully resourced.
- **Uncommitted**: The activities are planned, but not included in the budget, because either they are years in the future or they require a "go" decision based on a trigger (many companies call this a "buy-up" investment opportunity).

This is demonstrated in the following figure (Figure 11.2):

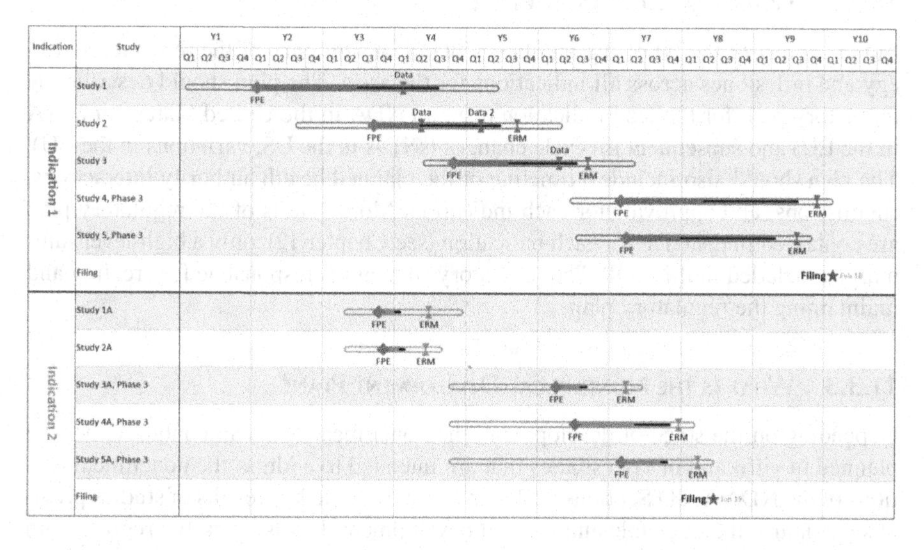

FIGURE 11.1 Example roadmap that can be used to show the trials contained in an asset development plan.

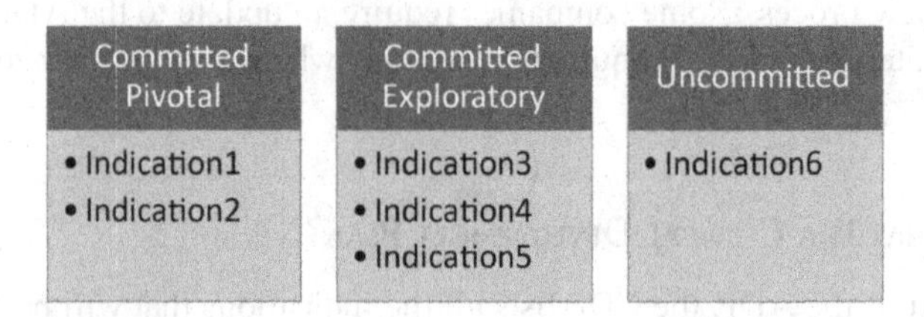

FIGURE 11.2 Skeleton of a slide that can be used to summarize the development strategy for an asset development plan.

For each indication, I will include a few slides from the CDP that provide the scope and investment commitments for each indication. The table below summarizes the contents of each category. There are templates for the target product profile, study design schema, and governance summary in Chapter 12.

Committed Pivotal	Committed Exploratory	Uncommitted
• Target product profile • Study design schema for each study • Regulatory path • Governance summary for each study	• Study design schema for each study • Governance summary for each study	• Target opportunity profile • Study design schema for each study • Business case summary

11.3.2 What Is a Regulatory Plan?

In the context of the ADP, the regulatory plan is a summary of the regulatory strategy and milestones across all indications for the asset. The plan should describe the regulatory path for the lead indication (e.g., an NDA in the United States, an MAA in the EU) and subsequent lifecycle changes (sNDA in the US, variations in the EU). The plan should also include a timeline of the planned health authority interactions, submissions, and approvals for each indication. More details of the regulatory plan are contained in the CDP for each indication (see Chapter 12); only a high-level summary is included in the ADP. The regulatory subteam is responsible for creating and maintaining the regulatory plan.

11.3.3 What Is the Nonclinical Development Plan?

Depending on the stage of development, the nonclinical plan can contain either the planned in vitro and in vivo studies that are intended to address the nonclinical sections of the NDA (CTD Sections 11.2.4 and 11.2.6) or the key results of studies previously conducted that enable and support development. If it is an early program with little or no preclinical data, the nonclinical plan should describe the choice of animal species and the rationale for the doses chosen for preclinical studies. If preclinical data exist, this section should describe any unusual or unexpected findings and how they will be addressed in the clinical program, such as results from an in vitro drug metabolism study that warrants further evaluation in an in vivo drug–drug interaction study. The nonclinical subteam is responsible for creating and executing the nonclinical plan.

11.3.4 What Is a CMC Plan?

A CMC plan outlines the activities that will be taken to develop, analyze, and manufacture drug substances and drug products for each phase of the program. The CMC subteam is responsible for creating and executing the CMC plan.

The CMC plan may include the following topics:

- Strategic assumptions for the program with respect to formulations that will be used for each stage and when each formulation will be introduced to the clinic
- The approach to address the specifications in the quality target product profile (TPP) (see Chapter 12) and the good manufacturing practices (GMP) regulatory requirements for each formulation that will be used in the clinic
- Sourcing strategy for raw materials and regulatory starting materials
- An assessment of the COGs for the drug substance and drug product
- The manufacturing schedule that will support drug supply needs during clinical development and commercialization. I typically see this schedule displayed in a Gantt with three swimlanes: drug substance, drug product, and analytical methods
- The packaging, labeling, and shipping plan
- The plan for obtaining stability data for regulatory submissions
- List of planned CMC-specific health authority interactions
- The risk management approach that will be used, such as failure mode and effects analysis (FMEA)

11.3.5 Medical Affairs Plan

While the CDP outlines the activities the R&D function will undertake to provide data to support regulatory approval, the medical affairs plan outlines the activities that will generate data to answer questions that healthcare payers and providers may have to optimize patient benefit. These activities can be divided into prelaunch and postlaunch and can include the following:

- Nonregistrational data generation (NRDG)
- Investigator-sponsored research (ISR) studies
- Stakeholder development (key opinion leaders—KOLs, advocacy groups, professional organizations)
- Scientific communications, congresses, and publications
- Medical information
- External medical education.

Including this information in the ADP allows the project team to incorporate some elements into the development plan so that data can be available at the time of launch. For example, the NRDG and ISR activities can be included in the program's overall integrated evidence plan (IEP); KOLs and advocacy groups can be involved in recruiting patients into clinical trials; and publications and presentations at congresses can be used to make the medical community aware of the development program in case they have patients who needed the investigational therapy.

11.3.6 What Is a Commercialization Plan?

A commercialization plan outlines the activities that will be taken to launch and market an approved drug product. The commercial subteam is responsible for creating the commercialization plan, which is shared with and endorsed by the core team. By building the commercial strategy into the ADP early in the drug development process, project teams can identify market opportunities and critical success factors that might otherwise be overlooked. For example, the commercialization plan can highlight key differentiators from competitors and the types of data that can be used to capitalize on these differentiators. The development team can then build some of these aspects into the development plan so that data are available at the time of launch.

A commercialization plan may include the following:

- Market overview (epidemiological data and trends)
- Competitive intelligence analysis, including current and emerging therapies
- Positioning
- Market access plan
- Digital strategy
- Pricing and reimbursement
- Sales and revenue forecast
- Lifecycle management (territory expansion, indication expansion, collaboration opportunities).

Planning a commercial strategy begins with a comprehensive review and assessment of current and emerging therapies. The value of your newly approved drug can be described in terms of differentiation from these competitors with respect to effectiveness, tolerability, and patients' and prescribers' perceptions of existing therapies. Some of these aspects can be included in the development plan so that data are ready at the time of launch. For example, before granting reimbursement, insurers sometimes require evidence that a new drug provides patient or health economic benefits that offset a higher cost compared to existing drugs. This consideration can be included in the development program by adding patient-reported outcomes (PROs) and health-economic outcomes endpoints in the trial design.

The sales and revenue forecast is often used to estimate the economic value of the drug asset. As described in Chapter 5, these estimates are part of the portfolio management process for selecting and prioritizing programs. Therefore, it is important to include these numbers in the ADP so that the same set of assumptions can be used across the organization.

Beyond the initial launch, the commercialization plan should describe lifecycle management opportunities that extend and maximize the value of a newly marketed drug. This can include expanding into international markets (see Chapter 7), expanding the label into other diseases or conditions (i.e., line extension), and collaboration/partnering opportunities designed to reduce overhead costs while still bringing in revenue. By including this information in the ADP, the project team and management can understand the long-term commercial potential of the asset.

11.3.7 What Is a Product Protection Plan?

To maximize the value of an asset, project teams and management should continuously evaluate ways to protect a product's market share. This is done via patents and regulatory exclusivity that effectively control the timing of the asset's loss of exclusivity (LOE). While it is beyond the scope of this book to describe the complex process of patent protection, project managers should have a basic understanding of the way patents are granted so that he/she can ask the right questions of the patent attorney on the project team and then include the key information in the ADP.

Patents are a property right granted by a legislative authority (e.g., USPTO) that prevents competitors from exploiting an inventor's intellectual property for a defined period of time. There are several types of patents that can be used to protect pharmaceutical products, including a composition-of-matter patent on the active pharmaceutical ingredient itself, patents for key steps in the manufacturing process, and patents for unique formulations. Patentable discoveries can be made throughout drug development, and the strength of patent protection will result from the family of patents that covers a variety of aspects of the asset. The project manager should work with the team's patent attorney to outline the key patents and their terms so that the team and management have a clear understanding of the LOE.

Exclusivity refers to a health authority's (e.g., US FDA) agreement to delay approval and prohibit the sale of competitor drugs for a defined period of time. In the United States, this legal framework that governs the introduction of generic competition is called the Hatch-Waxman Act, and it was designed to promote a balance between new drug innovation and greater public access to drugs that result from generic drug competition. The period of exclusivity is granted at the time of market approval, and the duration of protection varies depending on the type of exclusivity (e.g., new chemical entities via NDAs are granted 7 years of exclusivity; patent challenges via ANDAs are granted 180 days of exclusivity).

In the ADP, a single slide diagramming the key patent claims and exclusivity protections and the resulting timing of the LOE is usually all that is needed. The project manager can work with the team's patent attorney to obtain this information. The patent attorney should also be available to the project team for ideas and questions about potential patent and regulatory exclusivity opportunities.

11.4 SUMMARY

An ADP is the project team's guide for how to maximize the value of an asset. Value can come in terms of generating revenue from sales to patient populations as well as reducing costs to produce the product. The ADP serves to align with senior leadership on the strategic objectives of the asset, as well as to align the project team on the scope and timing of major work streams. The ADP is a living document, and it should be used regularly and kept up to date with current information.

12 Creating a Clinical Development Plan

Joseph P. Stalder
Groundswell Pharma Consulting

CONTENTS

12.1 INTRODUCTION

Whereas the Asset Development Plan is a high-level plan for how to maximize the value of an asset, the Clinical Development Plan is a more detailed plan that lays out the path for obtaining regulatory approval of a single indication. Depending on the therapeutic area, there may be several CDPs per ADP. This chapter will describe the utility of the CDP, the contents of a CDP, and a process the project manager can follow to create a CDP.

In principle, the CDP tells the story of how a drug candidate will be developed for a specific indication. The story includes the rationale for development (also known as "the reason to believe"), the investigational plan and decision points that will be used to determine if the drug candidate will satisfy regulatory and commercial expectations, and the resources required to execute the plan. This story is used to convince internal governing bodies that the project is worth pursuing, in essence saying,

DOI: 10.1201/9781003226857-14

"If you give us the resources we're asking for, we will deliver information to support the next decision point". The CDP, then, becomes the documentation of aligned expectations between the project team and the governing body.

In practice, the CDP outlines the approach for generating the data to include in the clinical sections of a market authorization application (i.e., Modules 2.5 and 2.7). Once approved by an internal governing body, it then empowers the project team with freedom to operate within the parameters defined within the CDP, making for a faster project execution. Furthermore, the CDP becomes the blueprint for the clinical sub-team (CST) to design and execute clinical trials. The CST is responsible for ensuring that every protocol contributes to the body of knowledge specified in the CDP and that the resulting data will enable a decision at the subsequent stage gate.

After approval at the appropriate governing body, certain slides from the CDP will be transferred to the ADP to keep the ADP up to date. The CDP, then, can be viewed as the mechanism for obtaining governance approval of the proposed plan, and the ADP becomes the record of the plan in the context of the rest of the development opportunities.

12.2 WHAT IS IN A CDP?

As stated previously, a CDP tells a story, and it is helpful to consider the storyline when building the CDP. The following outline is a good starting point for the story, but each CDP should be customized to the project's specific situation.

Topic Area	Topic
Commercial Rationale	• Disease epidemiology • Competitive landscape (current and emerging therapies) • Unmet need (medical, economic, accessibility) • Target markets and addressable market size
Scientific Rationale	• Mechanism of action • Summary of preclinical data [animal pharmacology (PK & PD), safety pharmacology, toxicology] • Summary of available clinical data (efficacy data from previously conducted clinical trials in the same indication or safety data from other indications)
Investigational Plan	• Target Product Profile • Roadmap of planned clinical trial(s) • Study design schema(s) • Statistical assumptions
Regulatory Plan	• Regulatory path to approval for all markets of interest • Regulatory intelligence (precedent, trends) • Summary of planned evidence to support approval • Timeline health authority interactions and submissions • Abbreviated regulatory target product profile focusing on potential areas of regulatory concern
Valuation	• Calculations of potential economic value (NPV, PYS, IRR, ROI, RRR)
Governance Summary	• Resource requirements (budget to next-stage gate/total project costs, FTEs, drug supply) • Go/no-go decision points • Risk assessment (probability of technical success and probability of regulatory success, if applicable) • Valuation (if applicable)

The sections below will describe in more detail some of the key components of the CDP.

12.3 HOW DOES A PROJECT MANAGER GO ABOUT CREATING A CDP?

A CDP is typically created by members of the CST. The format can vary from company to company, but I find that most companies now prefer a slide format because it makes the content easy to present to a governing body and the slides are often reused for other settings (e.g., the same study design schema slide can be used in the investigator training deck). For these reasons, I recommend building your CDP in a slide format and for the project manager to start the deck by creating a skeleton presentation of slides based on the table of contents above.

The project manager will coordinate the development of the CDP by assigning accountable persons to create certain sections of the document.

- The Commercial representative will populate the Commercial Rationale.
- The Nonclinical and Clinical representatives to populate the Scientific Rationale sections.
- The Regulatory representative is responsible for populating the Regulatory Considerations section.

The areas the project manager will need to focus most of his/her efforts are in the Investigational Plan and Strategic Assumptions sections, as these usually require cross-functional input and strategic thinking. I typically assume 6–8 weeks to create a fully fleshed out and vetted CDP that is ready for governance review. To populate the Investigational Plan and Strategic Assumptions sections, I find the following stepwise process to be most effective:

1. Define the Target Product Profile
2. Design the trial or trials that will provide evidence to support the target product profile (TPP)
3. Determine the operating assumptions and resources required to execute the trial or trial(s)
4. Evaluate the risks associated with the trial(s) and the indication
5. Summarize the key assumptions into a "Governance Summary"

12.3.1 The Target Product Profile

The first step for the project manager in developing the Investigational Plan is to facilitate the cross-functional creation of the TPP. In general, a **TPP** represents the desired characteristics of a product that, if achieved, would make the product successful. In practice, the "product" can have different meanings depending on the audience and on the stage of development, and thus, there can be many types of TPP in biopharma (see Sidebar). In terms of the CDP, we are interested in the Development TPP (dTPP).

12.3.1.1 What Is a dTPP and Why Should a Project Manager Care?

A dTPP defines the desired safety, efficacy, and patient-reported outcome characteristics of a drug that will make it commercially differentiated on the market for a particular indication. Commercial differentiation is benchmarked against a current standard of care, and the drug product is given a minimal, base, and optimal profile to differentiate the drug against the standard of care. An example of a dTPP is provided below:

Indication:				
			Product Profile	
Attribute		**Standard of Care**	**Base**	**Optimal**
Efficacy	1° Endpoint			
	2° Endpoints			
Safety	AEs/SAEs			
	Discontinuations			
PROs				

The *minimal* profile represents the characteristics needed not only to obtain regulatory approval but also to be competitive with the standard of care. This is typically equal to or slightly more desirable than the standard of care for at least one aspect of the drug product. The *optimal* profile represents the characteristics that would enable swift adoption and market dominance. It is the most highly differentiated drug product that prescribers, payors, and patients would want to see. The *base* profile represents the characteristics that are most likely to be achieved. This profile typically has several differentiating factors compared to the standard of care.

The dTPP is a very useful tool for the development project manager for many reasons, including the following:

- It helps build consensus on the definition of project success
- It is used to define the scope of the investigational plan
- It serves as a tool and framework for stage-gate decision-making
- It provides a set of assumptions for estimating the project's probability of success

If built correctly, the dTPP becomes the team's definition of "what does success look like". When I say "built correctly", I mean the dTPP is created from cross-functional input and collaborative negotiation. More on this in Section 12.3.1.2.

Once created, the dTPP can be used to create the Investigational Plan of the CDP. The base profile in the dTPP should be used to guide the design, conduct, and analysis of trials, as well as to ensure that there is adequate collection of any additional information desired to support commercialization (e.g., patient-reported outcomes). Specifically, the target safety and efficacy thresholds defined in the dTPP can be used to define the objectives and powering of the trial. For example, a standard-of-care oncology regimen may have an overall response rate (ORR) of 60%, so the Commercial team wants the target efficacy to be at least 70%. The Development team then will need to design a study that is powered to achieve a 70% ORR.

After trial results are available, the dTPP is useful for defining the go/no-go criteria for the next stage. Using the example above, if the efficacy result hits the 70% target and all other aspects of the dTPP are met, the decision to move forward should be clear. It becomes less clear when some but not all thresholds are achieved, and this is where senior leaders in the organization will need to apply expert judgement on the decision. If the minimal profile is not met, the dTPP provides support for the decision to terminate the program.

When project teams estimate the probability of success, they're really predicting the likelihood of the results of the Investigational Plan being able to meet the base case of the dTPP. Section 12.3.4 describes in more detail the use of the dTPP in estimating the probability of technical success (PTS) and probability of regulatory success (PRS).

THE VARIOUS TYPES OF TPP

There are several types of TPP (Research, Development, Marketing, Quality, and Regulatory), and each has its place in defining a specific desired outcome. We will touch briefly on Research, Quality, Regulatory, and Development TPPs in the sections below. We'll leave the Marketing TPP out of scope because this is usually created within the Commercial team for market research purposes and rarely does the Development Project Manager get involved.

It is important to remember that a TPP is more than just a document. It is the output of a process that encourages transparent communication across functions. While creating and updating the TPP, all team members participate in sharing their function's vision, expressing their function's needs and dependencies, and defining overall project expectations. In doing so, the process can reveal misalignments, gaps, blind spots, and uncertainties, thereby flagging areas that the Project Manager will need to address through strategic discussions with the project team and senior management.

Research TPP

A **Research TPP** (rTPP) is useful in drug discovery to define the desired pharmaceutical and pharmacological properties of a drug to put forward in clinical development. Note that this TPP is at the asset level, not the indication level. The Research Project Team is responsible for creating rTPP, usually at the point when a Project Manager is assigned to the team (see Chapter 2). Some examples of parameters that are included in an rTPP may be:

- Aqueous solubility
- Selectivity
- Blood–brain barrier penetration
- Half-life
- HERG activity
- Screening Ames
- Safety screening
- Drug–drug interactions (e.g., CYP inhibition/ induction)

There may be indication-specific parameters in the rTPP. For example, in oncology, the rTPP may include cellular activity and selectivity over wildtype values. For infectious diseases, the rTPP may include a spectrum of antimicrobial activity.

Quality TPP

A **Quality TPP** (qTPP), as defined in ICH Q8(R2), "forms the basis of design for the development of the product". Note that this TPP is at the asset level, not indication level. Considerations for the qTPP could include:

- Intended use in clinical setting, route of administration, dosage form, delivery systems
- Dosage strength(s)
- Container closure system
- Therapeutic moiety release or delivery and attributes affecting pharmacokinetic characteristics (e.g., dissolution, aerodynamic performance) appropriate to the drug product dosage form being developed
- Drug product quality criteria (e.g., sterility, purity, stability, and drug release) appropriate for the intended marketed product

Regulatory TPP

A **Regulatory TPP** (RegTPP), as defined in 2007 FDA Guidance *Target Product Profile – A Strategic Development Process Tool*, provides a format for discussions between a sponsor and the FDA that can be used throughout the drug development process. The RegTPP is sometimes called a Draft Label or Annotated Label because it contains all the elements of the structured product label. Indeed, as defined by the FDA Guidance, the TPP is "a format for a summary of a drug development program described in terms of labeling concepts". RegTPPs are especially useful for communicating with regulators when a product is novel and there is little experience over the entire development lifecycle, such as cellular and gene therapies.

Bringing back the concept of "keeping the end in mind", the RegTPP is the representation of the goal of an indication-specific product development plan – the product label. The FDA Guidance has a nice template for the RegTPP, so I will not duplicate it here, but instead list the sections that are included in the template:

- Indications and Usage
- Dosage and Administration
- Dosage Forms and Strengths
- Contraindications
- Warnings and Precautions
- Adverse Reactions
- Drug Interactions
- Use in Specific Populations
- Drug Abuse and Dependence

- Overdosage
- Description
- Clinical Pharmacology
- Nonclinical Toxicology
- Clinical Studies
- References
- How Supplied/Storage and Handling
- Patient Counseling Information

For each of these sections, the sponsor creates three fields: Target, Annotations, and Comments. The Target field includes intended labeling language based on the outcomes of the indicated studies. The Annotation field includes planned or completed studies to support the proposed claim, including protocol numbers, submission numbers, and dates. The Comment field contains additional information that can facilitate communication between the sponsor and regulators.

In my experience, the Regulatory Team typically creates the RegTPP, with support from the project manager to resolve conflicting opinions at the development core team level. The Regulatory Team then uses the RegTPP for interactions with regulators as needed [e.g., when submitting a request for a product name to European Medicines Agency (EMA) (Invented) Name Review Group].

Development TPP

A **Development TPP** (dTPP) defines the desired characteristics of a drug that will make it commercially differentiated on the market. Similar to RegTPP, a dTPP is specific to an indication. However, in contrast to the RegTPP, the dTPP is an internal document that represents an agreement between Commercial and Development on what the market will want and what can be achieved in clinical trials to support desired marketing claims. Thus, it becomes the "requirement document" for the CDP that defines the differentiating aspects that will make a drug commercially attractive for a given indication.

12.3.1.2 How Does a Project Manager Go about Creating a dTPP?

As described in Chapter 15, good project management begins with setting goals because it forces the team to begin with the end in mind. Consider, then, the dTPP as part of the project's goals, along with the milestones and deliverable schedule that the project manager creates. It makes sense, then, that the best time to create a dTPP is when the initial plan is being defined. The team can then use the dTPP to plan the work that needs to be done to satisfy the dTPP, thereby ensuring that the work is relevant to the scope of the project. With respect to the CDP, by first creating the dTPP, the CST can then design an Investigational Plan that can provide the safety and efficacy evidence needed to convince regulators, payers, providers, and patients.

I have seen several companies that generate the dTPP within Commercial and then share it with the core team as a final output. To me, this is like providing a scent rag to the core team and telling them to "go get it!". Instead, the dTPP should be developed collaboratively by all stakeholders in the project team. This then puts the development project manager in a position to guide the project team through creation of the dTPP, with inputs from Clinical, Commercial, Biostatistics, Regulatory, Medical Affairs, Safety, Pricing and Market Access, HEOR, and others. The result is not only a thoroughly vetted set of requirements but also a team that is aligned and supportive of the dTPP.

As drivers for the creation of the dTPP, the project manager should become familiar with his/her organization's dTPP options. Each company may have tailored dTPP formats that are matched to the organization's maturity and the project's stage of development. While there is no universally accepted format for a dTPP, the guiding principle for content to include in the dTPP is the differentiating factors that will make the product commercially successful. This can include indications or disease subsets to be treated, patient population or subpopulation, clinical safety or efficacy, formulations, dosing regimen, route of administration, drug–drug interactions, or contraindications and precautions.

If you are introducing your team or company to a dTPP for the first time, here are two tips:

1. start simple
2. make sure that everyone is aware of how the TPP should be used.

I recommend the simple 1-pager format below as a starting point and only building one dTPP. Senior leaders sometimes get carried away with the number of TPPs that can be developed with different thresholds for each category (primary endpoint, secondary endpoints, safety parameters, patient-reported outcomes, etc.). Starting with just one dTPP and having minimal, target, and optimal thresholds usually provide enough context to design the right study and make the right go/no-go decision based on study results.

Make sure everyone is aware that the dTPP is a *decision-support tool*, not a decision-making algorithm by itself. Some leaders and team members get concerned that the project will be terminated if the study results do not meet the dTPP exactly. However, in practice, it is expert judgement from *people*, not the dTPP alone, that will determine the path forward for the project. The dTPP only provides a framework and context for the decision that needs to be made.

Because the dTPP is used for estimating the probability of success and for go/no-go decision-making, it is important to keep it updated with the most recent information. Thus, the dTPP should be considered a living document, and it is the project team's responsibly to update the requirements with new information when it becomes available. This new information could come from within the project as studies read out and new data become available or outside the project with respect to competitive intelligence or regulatory landscape, such as a change to the standard of care. When incorporating this internal and external information, the project team should consider the implications of new information on the project's probability of success, and if there is a significant change to the probability of success, a governance interaction may be needed to realign the project to the portfolio.

12.3.2 Investigational Plan and Trial Designs

This section of the CDP provides details on the studies that will provide the data to support the marketing application. The author likes to start with a high-level roadmap of the investigational plan such as the one below, which uses the OnePager software to visualize project information in a simple Gantt chart. The project manager is responsible for generating this slide. The roadmap helps the team and governing bodies to visualize various workstreams that will take place throughout the rest of the program. It supports an understanding of the dependencies and coordinated activities, triggering of financial and resource commitments (e.g., drug manufacturing and scale-up, formulation development, toxicology and drug metabolism studies, or validation of biomarkers and/or companion diagnostics, or seeking regulatory agency advice), and timing of major milestones that serve as catalysts for the organization (Figure 12.1).

As you can see in the above Gantt, only a few major milestones are included: First Patient Enrolled, Data Readout, and Early Release Memo (aka Top-Line Report). This keeps the focus on the major inflection points that an Executive Committee would be interested in knowing. Also notice that the investigational plan contains all the studies needed to submit a market approval application for a single indication. Though the "ask" to the EC may only be to fund a certain study or set of studies to get to the next stage-gate, it is helpful for the governing body to see the entire development path and know the total development cost to get the desired indication.

Another example of an investigational plan that has a bit more detail, also from OnePager, is shown below. This one nicely calls out the stage gates that represent the investment decision points along the development path as well as the types of studies that are included in each phase (Figure 12.2).

After presenting the development roadmap, I like to dive deeper into the study details. A study schema slide like the following is helpful to give the high-level summary of the trial population, design, and endpoints. The Clinical representative is responsible for generating this slide (Figure 12.3).

It is also helpful to have a slide that describes the statistical assumptions that were used to calculate the sample size. Some key assumptions may include the

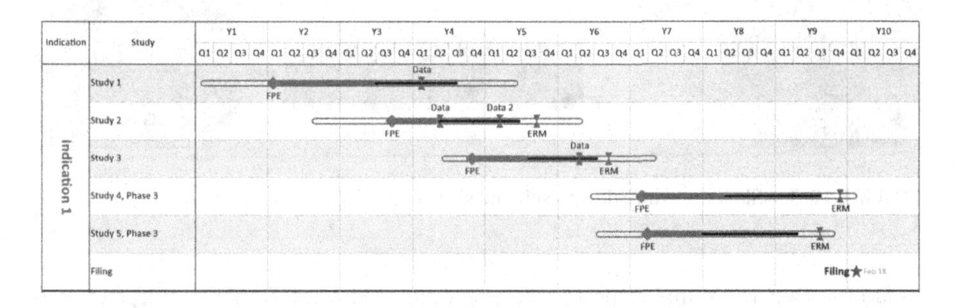

FIGURE 12.1 Example roadmap that can be used to show the trials contained in a clinical development plan.

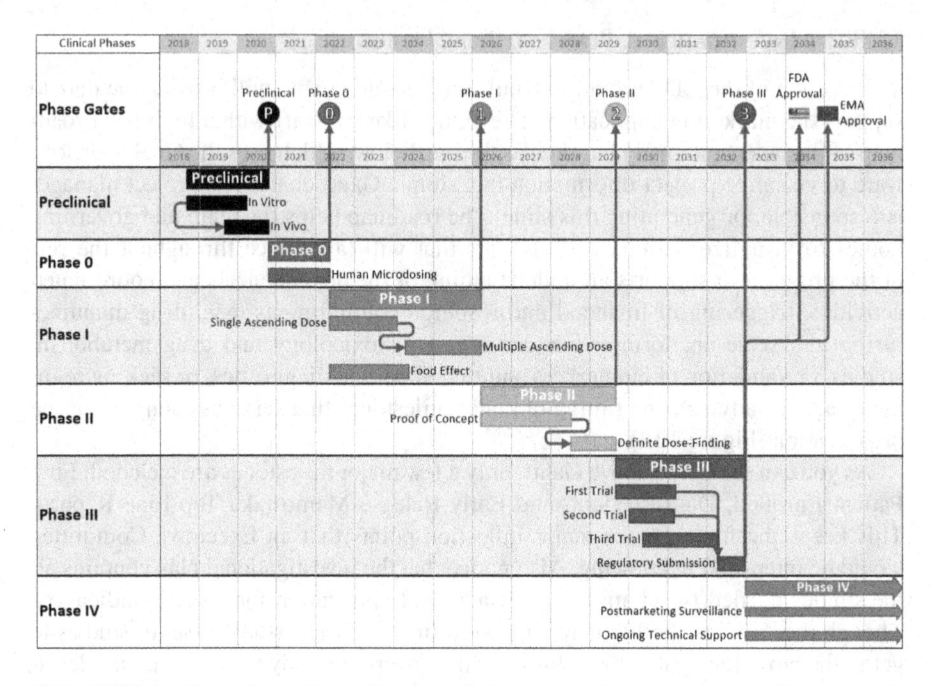

FIGURE 12.2 Example roadmap showing all the trials and stage gates for a clinical development plan.

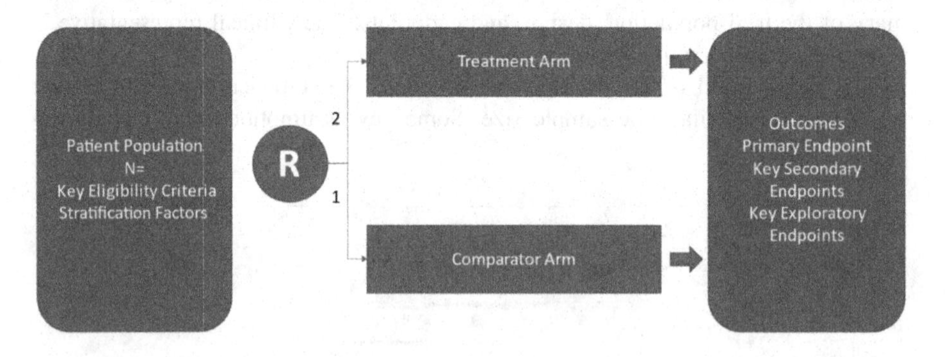

FIGURE 12.3 Skeleton study design schema slide.

randomization ratio, alpha, power, hazard ratio, and total N. This slide can also include scenarios that were considered but not selected by the team, just to show the full scope of the assessment the team did when evaluating study designs. The Biostatistician is responsible for generating this slide.

12.3.3 Regulatory Plan

In the context of the CDP, the Regulatory Plan outlines the regulatory path to approval for the indication. The regulatory path may differ among health authorities (e.g., FDA, EMA, PMDA, and NMPA), and the Regulatory Plan should highlight these differences and the unique requirements for each agency. Each regulatory requirement should be juxtaposed with a summary of the planned evidence to address it, as well as guidelines and precedents that support the proposed evidence-generation plan. Alternative approval pathways may be available (such as accelerated approval), and the plan should note these alternatives and the rationale for the recommended approach. The plan should also describe opportunities for special designation (e.g., fast-track, breakthrough, and orphan disease indication status).

The Regulatory Plan should include a timeline of key health authority interactions (e.g., pre-IND, end-of-phase meetings, and SAWP meetings) and major regulatory submissions (e.g., IND, NDA/BLA, and MAA), overlaid onto the program roadmap that includes trial milestones such as First Patient Dosed and Data Readout and other events that may trigger regulatory activities such as introduction of a new dosage form that would require a Chemistry, Manufacturing, and Controls (CMC) submission.

The Regulatory Plan may include an abbreviated rTPP, focusing on areas of potential regulatory concern that will need to be addressed in the CDP through nonclinical studies, clinical trials, and CMC activities. These areas of concern can be further explained in the plan for seeking health authority feedback in the form of questions and objectives for health authority interactions.

12.3.4 Operating Assumptions and Resource Requirements

The project manager, with input from Clinical Operations and Finance, is heavily involved in creating this section of the CDP because it relates to scheduling, budgeting, and resourcing the project. The following elements may be included in this section:

- Study Timeline with milestones such as final approved protocol, first site activated, first patient in, last patient in, last patient out, database lock, top-line report, final tables/listings/and figures, and final clinical study report
- Study assumptions (number of sites/ regions/countries, enrollment rate) and strength of assumption (feasibility data)
- Delivery model (insourced and outsourced) and resource demand assessment
- Development cost, including investment to next decision point and remaining cost to launch

The template table below can be used to show the projected development cost and resource demand by year. Depending on your organization's stage gates, the sum costs may differ. In the example below, the team is requesting funds to go from preclinical to proof of concept and giving line of sight to potential costs to pivotal data readout that would support market application submission. The Total Costs row provides an estimate of the cost to launch. By time-phasing the cost estimates, governing bodies can see when spending will occur, giving them the opportunity to accelerate or decelerate activities to match the budget if needed.

Phase ($M)	Y1	Y2	Y3	Y4	Y4	Y5	Y6	Y7	Total
Preclinical									
Phase 1									
Phase 2									
Sum Cost to Proof of Concept									
Phase 3 – Study1									
Phase 3 – Study2									
Sum Cost to Pivotal Readout									
Total Costs									

12.3.5 Risk Assessment (Probability of Technical and Regulatory Success)

Risk assessments in the CDP are defined in terms of the likelihood of progressing through the major stages of development. PTS assesses the likelihood that the clinical development program will meet the predefined characteristics of the base case dTPP. PRS is the likelihood that, having achieved technical success (i.e., met the base case dTPP), a market application will be approved. Thus, each phase transition (e.g., Phase 1 to Phase 2) carries its own probability of success, and when multiplied across all phases and with the likelihood of regulatory approval, the project team can estimate an overall probability of technical and regulatory success (PTRS). The PTRS is sometimes called the Probability of Launch or the Likelihood of Approval.

12.3.5.1 Why Should a Project Manager Care about the PTRS?

The PTRS is a critical factor for project management and portfolio management. Although there is no way for project managers to completely mitigate these risks (as described in more detail in Chapter 16), the process of identifying and assessing risks allows the team to be aware of and agree to the risks so that there are no surprises or disappointments if things do not turn out as desired (e.g., a clinical trial readout failing to meet the primary endpoint).

For portfolio management, the PTRS is used in many ways:

- To calibrate ongoing projects and investment opportunities for portfolio reviews intended to make strategic decisions on which projects to accelerate, decelerate, or terminate
- To adjust the valuation estimates [e.g., NPV (net present value)] that are used in the long-range plan
- To predict the project volume so that appropriate budget and resource capacity can be planned

An example of an output of the PTRS exercise is shown below. This is for a project in the ophthalmology space, but the concepts apply to any therapeutic area. The table describes the likelihood of transitioning from one phase to the next. The benchmark probability of success (POS) is obtained from the published literature that reports on industry averages. The proposed POS is the estimate the team assigned after up- and down-modulating the benchmark for the project-specific factors. The rationale describes the reason the proposed POS differs from the benchmark POS.

Phase	Benchmark POS (%)	Proposed POS (%)	Rationale
Pre-IND	90	100	Phase complete
Phase 1	40	100	Phase complete
Phase 2	70	60	Efficacy endpoint used for Ph2 is different from the one used in Ph1
Phase 3	60	50	Comparator performance is inconsistent, and if it reads out at the higher end of the range, our Ph3 may not meet statistical significance
Approval	80	80	No change from industry average
Overall	12	24	

The value of going through this PTRS exercise is less about the resultant PTRS number (the number is always wrong anyway) and more about the process of identifying and assessing the risks and then using the PTRS to calibrate the project's value against the portfolio of other development opportunities. This is where a governing body needs to be involved, as their remit is to assess the team's proposed PTRS against other development plans that have been proposed.

12.3.5.2 How Is PTS Estimated?

The experimental nature of clinical trials requires us to design a study using several assumptions about how the investigational product will behave in a population of people compared to another treatment. Even when information from prior studies is available, there are several factors that contribute to the uncertainty of the outcome of the study. This risk is represented by a PTS, which assesses the likelihood that the clinical trial will meet the predefined technical criteria (e.g., pharmacokinetics, safety, tolerability, and efficacy) at the final analysis to satisfy the key elements of the base case TPP.

As an example, let's look at the determination of a trial's sample size, a foundational component of trial design. Statisticians do their best to power a study by including enough study subjects to be able to detect a statistically significant difference in treatment effect between the investigational product and the comparator. However, this estimate is largely dependent on a "best-guess" assumption of treatment effect that physicians think is reasonable. In failed clinical trials, this assumption is often found to be wrong. Even when a placebo is used as the comparator arm, the placebo group can do better than expected, thereby making the difference in treatment effects not statistically significant.

The table below provides some additional considerations when assessing the PTS.

Strength of Clinical Rationale	• Are the existing clinical data consistent with the biological hypothesis? • Are all existing clinical data consistent? • Are the data from primary and supportive endpoints consistent? • Can all findings (especially if any discrepancies) be explained? • Does the method of handling missing data affect the conclusions made? • Are subgroup findings plausible? • Is the sample size sufficient to extrapolate results? • Is there value in incorporating prior belief in subgroup by way of a Bayesian analysis? • How do results fit in relation to competitor data? • How robust is the existing data? Are there potential sources of bias? (e.g., randomized? double-blind? Blinded independent central review? Validated endpoint?) • Any key information missing?
Strength of Pharmacological Rationale	• What is the level of confidence in the selected dose? • Have the doses studied allowed a full assessment of the dose-response relationship? • Is there a consistent dose response? • Has dose-response modeling been carried out? • How strong is the link between PK/PD modeling and clinical efficacy? Can this be used to understand how exposure relates to outcomes and tolerability?
Uncertainty due to differences in design vs previous trials	• Endpoints (e.g., definition, validation, timing, and maturity) • Patient population (i.e., inclusion/exclusion criteria) • Sample size (assuming larger variability, is the treatment effect sufficiently discounted?) • Background therapy • Comparator • Geographical representation and multicenter variability • Is the same formulation being used from Ph2 to Ph3? • Is there a need to change formulations during the pivotal trial?

12.3.5.3 How Is PRS Estimated?

PRS is the likelihood that, having achieved technical success by meeting the base case TPP, a market application will be approved for the indication being sought. There are a number of considerations for the team to assess when evaluating the PRS, including the following:

- The strength of the clinical data package
  - Trial design, including comparator used
  - Trial endpoints
  - Number of trials
  - Statistical considerations
- Completeness of supportive data packages
  - Nonclinical data package
  - CMC data package

- Early signals of HA opinion based on prior interactions
  - Has Scientific Advice been sought from relevant HAs? If so, has the advice provided by HAs been followed?
  - Has the program received any expedited review designations from HAs (e.g., Fast-Track, Breakthrough Therapy, Priority Review, RMAT, PRIME, Sakigake)?
  - Are there any recent regulatory precedents, known outcomes of HA assessment, or known decisions on similar programs?
  - Potential for regulatory flexibility due to high unmet medical need

When estimating the PRS, the Regulatory Team usually comes up with the first proposal for the Core Team to review. Typically, they will start with a benchmark and then modulate up and down based on strengths and weaknesses in the strategy and planned dataset. Upward adjustments are applied based on factors that improve the potential for approval; downward adjustments are applied based on factors that reduce the potential for approval.

Potential upward adjustments include:

- Having prior HA input on the program through expedited review designation, Special Protocol Assessment, or other special designations (e.g., orphan drug status, emergency use authorization)
- Degree to which the program addresses an unmet medical need in the treatment of a serious or life-threatening condition, especially if the targeted condition is on a health authority list of critical therapies
- The need for an advisory committee (US)
- Compliance with health authority advice
- Compliance with health authority data requests/analyses
- Robustness of the planned clinical dataset
- Differentiating factors that improve patient experience and outcomes (e.g., a more favorable safety and tolerability profile compared currently marketed products)

Potential downward adjustments include:

- Noncompliance with HA advice
- Unmitigated technical risk or expected magnitude of benefit is not viewed as clinically meaningful
- Safety signals cannot be mitigated
- Unclear or unknown mechanism of action or therapeutic area science
- Efficacy is not comparable to current standard of care or a change in the competitive landscape is imminent that will make the comparator irrelevant
- The need for an advisory committee (US)
- Issues with compliance or inspection readiness

12.3.5.4 What Other Risks Should Be Considered in the CDP?

Some risks are inherent to the molecular structure of the asset or the fundamental mechanism of action that the asset exploits to treat the targeted disease. If the asset exhibits some characteristics that are unacceptable, there is very little that can be done to work around the problem. For example, from a purely chemical perspective, a drug may have risks associated with manufacturability and scalability that cannot be overcome. Because not every step in a process can be sufficiently scaled up, upscaling may result in reduced yields, scale-up may introduce unacceptable impurity levels, or the drug may have unacceptably short stability or meticulous storage conditions.

An asset may have undesirable therapeutic effects that cannot be managed, such as the "class effect" that we commonly see in the label as boxed warnings. For example, the cardiovascular risk associated with COX-2 selective nonsteroidal anti-inflammatory drugs or tendon rupture associated with fluoroquinolones have resulted in withdrawal of market approvals and boxed warnings intended to warn physicians and patients of the severity of the risk.

Another example of asset-level risks is the use of advanced therapies such as gene therapy, cell therapy, and tissue engineering. Being so new to the field, these treatment modalities carry a higher level of risk of getting to market compared to better-known small molecule or biologic treatments.

The following considerations are helpful when assessing asset-level risks:

- Are there any relevant data from other compounds in this class?
- Is there a precedent for the class or mechanism of action?
- Are there any potentially unscalable manufacturing steps or bottleneck intermediates?

12.3.6 Valuation

The potential economic value of a project is commonly used to compare investment opportunities across the portfolio. Each company will have its preferred valuation metrics, such as PYS, eNPV, IRR, ROI, and RRR, so the project manager should become familiar with the metric of choice to represent in the CDP. It is beyond the scope of this section to describe the method for calculating discounted cash flows and expected net present values (eNPVs), but suffice it to say that these metrics are commonly used and are relevant in the CDP for many mid- to large-biopharma companies. I will note here that I find valuations to be more applicable for later-stage projects and those with registrational intent. As described in Section 5.3, trying to calculate an eNPV for an early-stage asset has a number of challenges and often the result is a negative value that looks unattractive compared to late-stage projects. For this reason, I typically only include the valuation in the CDP for a late-stage asset.

As described in Chapter 5, the Portfolio Management Team will create the project's valuation using the project team's inputs on scope, launch timing, cost of development, risk, and sales forecast. In addition, sales projections, pricing, and market share assumptions are typically reviewed and endorsed in a separate forum before being included in the CDP. Therefore, the project manager will typically only need to coordinate with the Portfolio Team to ensure that this information is available in the CDP.

12.3.7 Governance Summary

I find it helpful to summarize the key assumptions of the investment proposal into a Governance Summary slide that contains the following information:

Project Attribute		Description/Value
TPP	Indication	e.g., First-line treatment for patients with disease X
	Key Success Criteria	e.g., % improvement of current SOC
Budget	Cost to Next Stage Gate	$Xm (Internal/External)
	Total Cost to Launch	$Xm (Internal/External)
Timeline	Stage Start	MMM YYYY
	Stage Finish	MMM YYYY
	Launch	MMM YYYY
Risks	PTS	X%
	PRS	Y%
Valuation	PYS	$Xm (YYYY)
	eNPV	$Xm

Drug development using the stage-gate process requires clear definitions of "Go/No-Go" criteria to support the decision to transition between stages. The CDP should provide specific "Go/No-Go" decision criteria. Often these criteria are set against the ability to meet the TPP, and agreement on the TPP characteristics that warrant "Go/No-Go" designation typically involves Clinical Leads, Medical Affairs, Commercial, Biostatistics, and other project team members, as well as senior management.

12.4 SUMMARY

A CDP tells the story of how a drug candidate will be developed for a specific indication. It forms the evidence that will be contained in Modules 2.5 and 2.7 of the Common Technical Document for a market application submission. When used with project teams and governance, it forms the agreed plan for how a drug will be developed and that value should be achieved if the project goes to plan. Some key elements of the CDP include the dTPP, the investigational plan and trial designs, the regulatory plan, operating assumptions and resource requirements, an assessment of the design risk (in the form of a probability of technical and regulatory success – PTRS), and a summary of the value of the indication. The CST is usually best positioned to create the CDP and ensure its execution, with the project manager and project lead gathering updates and managing issues at the Core Team.

13 Creating a Market Application Submission

Joseph P. Stalder
Groundswell Pharma Consulting

CONTENTS

13.1 INTRODUCTION

Before a new drug is introduced for sale to the public in a country or region, the safety, efficacy, and quality of the new chemical entity must be reviewed by the regulatory authority of that country or region. Therefore, when a drug innovator (also known as sponsor) has collected enough evidence of a new drug's safety, efficacy, and quality for regulatory review, the sponsor submits to an application for health authorities to approve the marketing and sale of the new drug to the public. Hence, a marketing application represents the culmination of the work conducted under an asset's CDP and CMC Development Plan.

Each country or region may call the marketing application submission something different. For example, in the United States, it is called a New Drug Application (NDA); in the EU, it is called a Market Authorization Application; and in Canada, it is called a New Drug Submission. For purposes of this chapter, we will call it a MAS.

DOI: 10.1201/9781003226857-15

The content required in an NDA may differ by country and region, complicating the development of a global dossier intended for multi-regional launch. Although the International Council for Harmonisation has attempted to harmonize the content for market applications via the Common Technical Document (CTD), each regulatory authority has unique requirements and considerations for reviewing a marketing application. This is especially true for Module 1 of the CTD, where each country has its own content, numbering system, and requirements. In addition, regulatory requirements in each country often change, and these changes could affect the submission of a marketing application. It takes a very astute Regulatory Affairs department to keep abreast of the global regulations.

In addition to differences in the market application content and requirements across countries and regions, there may be different types of applications within the same country or region. In the United States alone, there are several types of market application, depending on the type of the product that is being submitted, as depicted in Figure 13.1. For devices in particular, as described in Chapter 8, the regulatory path can drive the amount of evidence that needs to be summarized in an application. Therefore, the Regulatory Affairs department also needs to keep the project team aligned on the right type of application so that the right data can be included in the submission (Figure 13.1).

Now consider the complexity of these multiple types of applications across the differing contents, structures, and requirement by country or region, you can see why the regulatory team has its hands full with submission planning. Here is a brief list of some of the more common regulatory agencies when submitting a global dossier:

- USFDA (the USA),
- EMA (European Union)
- MHRA (the UK)
- TGA (Australia)
- Health Canada (Canada)
- NMPA (China)
- ANVISA (Brazil)
- PMDA (Japan)
- SWISSMEDIC (Switzerland)
- KFDA (Korea)

It is beyond the scope of this chapter to recommend regulatory strategies, but a development project manager should know some of the expedited pathways and collaborative regulatory reviews that are now available. For example, in the United States, there are four programs to facilitate and expedite development and review of new drugs to address unmet medical needs (see Chapter 1 for more detail on these):

- fast track designation
- breakthrough therapy designation
- accelerated approval
- priority review designation

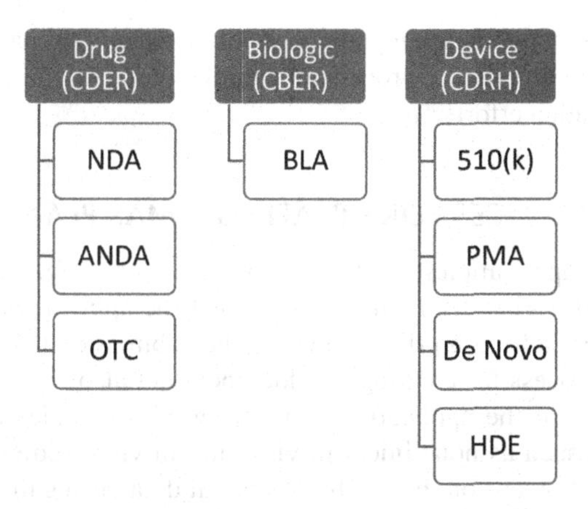

FIGURE 13.1 Types of market applications in the United States grouped by regulatory path.

In addition, there are collaborative review programs such as Project Orbis (Australia, Brazil, Canada, Israel, Singapore, Switzerland, and the UK) and the Access Consortium (Australia, Canada, Singapore, Switzerland, and the UK). These programs are intended to foster increased communication and collaboration among members to make new drugs available to patients faster and with less redundancy.

This chapter will focus on the development project manager's responsibility for contributing to the MAS Plan and the execution of the plan. Here we assume that there is no regulatory project manager to manage the submission process, and that the development project manager holds sole responsibility for coordinating the development of the MAS. We will cover the contents of a MAS Plan and walk through some best practices for coordinating the activities to create a MAS.

13.2 WHAT IS A MAS PLAN AND WHY SHOULD A PROJECT MANAGER CARE?

A MAS Plan outlines the process to prepare a marketing application document. It describes the activities, roles and responsibilities, timing, and dependencies that a submission team will follow to consolidate all the evidence to support market approval and write the narrative of a new drug's development journey.

At large companies, there is often a specialized group of people who support the development of major submissions. This group typically joins the project team several months before the final data readout to begin integrating with the team and establishing ways of working that will be used when all the data are available and the writing begins in earnest. There is often a regulatory project manager or a submission project manager who is assigned to this team.

In small- and medium-size biopharmaceutical companies, the development project manager is usually responsible for creating the MAS Plan and coordinating the activities to compile the submission. While the same project management skills are required to manage the team and coordinate efforts, the domain expertise of creating a submission may not be a development project manager's strength. This chapter is meant to outline a best practice process for creating the MAS Plan and managing the document preparation efforts.

13.3 BEST PRACTICES FOR CREATING A MAS PLAN

Preparation of an application should begin a year or more before the final data readout of the pivotal trial or trials that will support the application (termed top-line results (TLR)). By setting up the plan early, the submission team not only is able to align on the process for creating the document but also has time to proactively complete sections of the application that are based on studies that have already been conducted, such as nonclinical in vitro and in vivo studies. This allows the submission team to focus on the last bit of critical data during the final preparation of the application.

13.3.1 Before TLR

There are many activities that the team can perform early in the process to define how the submission will be compiled. Examples are given as follows:

- Define the content creation process, including the sequence of steps, the ownership of sections, and the author/reviewer assignment matrix.
- Prepare shell documents that define the level of granularity and table of contents for each module.
- Create a lexicon of common verbiage that authors can use to make content creation faster and more consistent.
- Create a detailed timeline for the development of each module, including the resources required for each task.
- Compile a list of "legacy" reports and discuss the need for errata, addenda, amendments, and meta-analyses of these reports.
- Consider risks to the plan.
    - Do you have enough people? Do you have the right people?
    - Are there any special analyses or reports that might delay the timeline (e.g., concentration QT analyses, population pharmacokinetics analyses, exposure–response analyses, and bioanalytical reports)

13.3.1.1 Team Structure

As with any project, a clear team structure will enable better coordination and communication to get work done. Before starting activities, a submission leader should be appointed. The submission leader and project manager will then outline a team structure and define the roles and responsibilities that will be needed to complete the submission. The submission leader is also responsible for aligning the team on expectations, securing the staff needed to complete the submission, and managing stakeholders throughout the submission process.

An example of a team structure is as follows:

- **Core**: Regulatory, Regulatory Operations, Clinical Science, Pharmacology, DMPK, Toxicology, Clinical Pharmacology, CMC, and Medical Writing
- **Extended**: IT, Legal, Corporate Communications, senior leadership, Clinical Operations, and QA
- **Sub-teams**: Clinical, Nonclinical, and CMC sub-teams can be formed to plan and deliver sections of the submission. These can operate as separate workstreams (e.g., CMC sub-team can work independently on Module 3), but eventually, they need to tie together around week T-15.

13.3.1.2 Author/Reviewer Assignment Matrix

With so many authors and reviewers involved in the process, it is helpful to assign people to roles as follows:

- describe accountable/ responsible parties for each section;
- start with functions and then add named individuals;
- populate the timeline and translate into the task manager; and
- once timeline is populated, archive the RAM to avoid discrepancies.

13.3.1.3 Defining the Content Creation Process

Content creation involves authoring and reviewing of modules in the MAS and close control of versions. Before authors begin writing, it is helpful to outline the process of content creation so that authors and reviewers know what to expect. This process can then be mapped into a Gantt chart with dependencies to create a schedule of activities. An example of an authoring process is shown in Figure 13.2.

This process is further explained in Figure 13.3. The importance of the medical writing function cannot be overemphasized here. Medical writers are able to convert the content from reports into the shell of the MAS, greatly easing the burden on the subject matter experts to create new verbiage. They also bring consistency to the story by adhering to the company's style guide, the document's lexicon, and the project's key messages (Figure 13.3).

13.3.1.4 Submission Content Plan

The MAS is a compilation of hundreds of documents that are developed over the life of the asset's development. These documents may be many years old and need updates in the form of errata, addenda, amendments, or meta-analyses. In order to keep track of all the documents that will be submitted, the team should create a submission content plan, or submission tracker, which lists the documents to be included, its location in the CTD, a document owner, a status, and a due date if revisions are needed.

The submission content plan needs should be up to date throughout the MAS development process. I found it useful to bring up the content plan with my sub-teams and

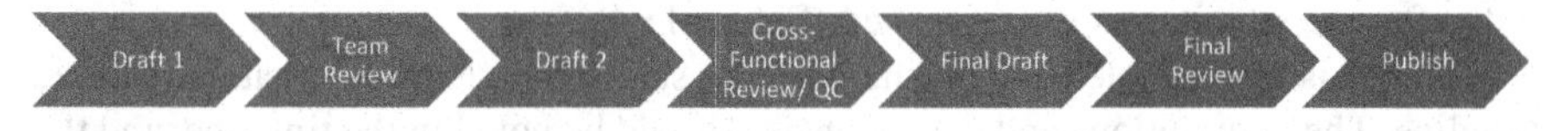

FIGURE 13.2 Example content creation process diagram.

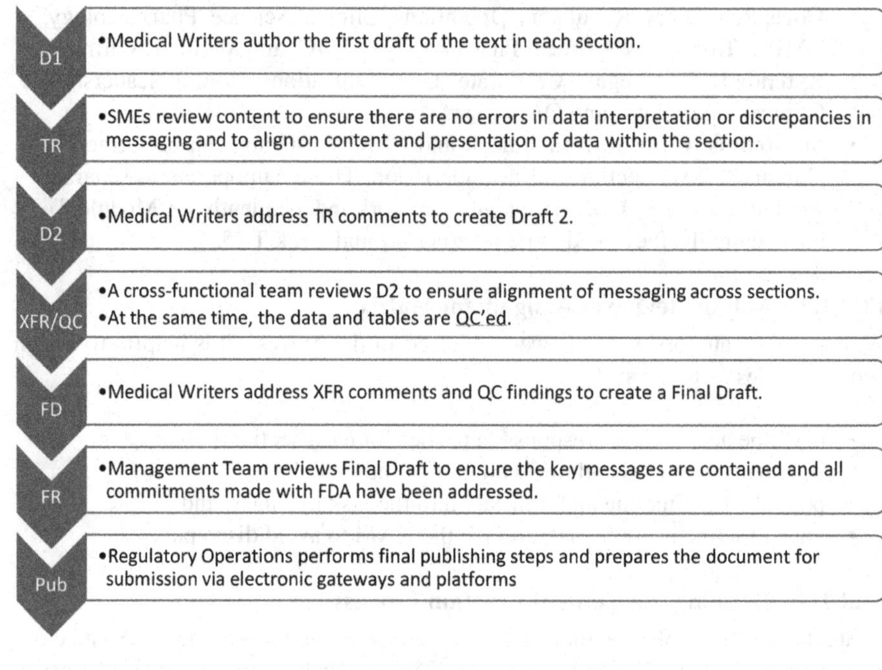

D1
•Medical Writers author the first draft of the text in each section.
TR
•SMEs review content to ensure there are no errors in data interpretation or discrepancies in messaging and to align on content and presentation of data within the section.
D2
•Medical Writers address TR comments to create Draft 2.
XFR/QC
•A cross-functional team reviews D2 to ensure alignment of messaging across sections.
•At the same time, the data and tables are QC'ed.
FD
•Medical Writers address XFR comments and QC findings to create a Final Draft.
FR
•Management Team reviews Final Draft to ensure the key messages are contained and all commitments made with FDA have been addressed.
Pub
•Regulatory Operations performs final publishing steps and prepares the document for submission via electronic gateways and platforms

FIGURE 13.3 Breakdown of roles and responsibilities for each step in the content creation process.

go through the status of each document on a weekly basis. Sometimes there would be dependencies between reports that needed to be managed and coordinated closely. Sometimes "surprise" documents would be discovered as authors read through existing reports, finding a reference to another report that they did not know already existed. Sometimes there would even be references to studies for which reports did not exist, so the functional subject matter experts would have to rush to create a new report. The submission content plan becomes a valuable tool to keep track of all these loose ends.

13.3.1.5 Project Roadmap

A project roadmap helps the team understand the "big picture" plan and how each step fits into the broader plan. I used these visualizations during the weekly standup meeting to indicate progress made and to ground the team on what is coming up. As shown in the example below for Module 2, all tasks related to authoring a section are grouped into a single workstream, so the handoff between responsible parties is clear. Note also that although the dependencies between sections are not indicated on the graphic, the link between sections can be inferred by the relative timing of each section. For example, authoring of the Nonclinical and Clinical Overviews began only after the summaries were finished (Figure 13.4).

The PM is responsible for working with the submission team to create a realistic timeline. The assumptions and dependencies should be noted in the timeline, and the plan should be adequately resource-loaded to avoid overloading certain functions

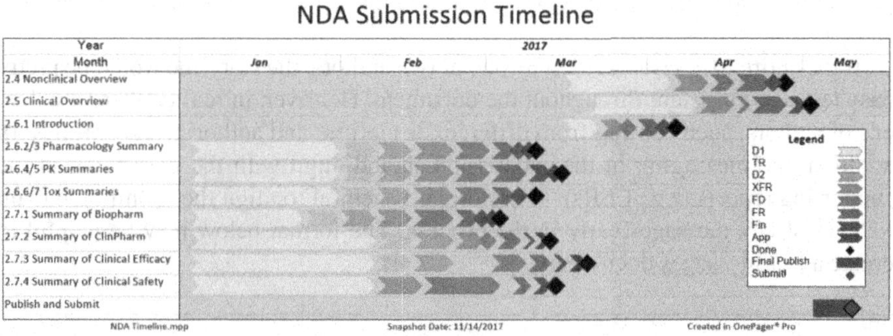

NDA Submission Timeline

FIGURE 13.4 Example of Gantt chart for a market application submission development project.

with spikes of activity. For example, when possible, QC steps should be staggered so that the QC department does not receive ten documents on the same day. Similarly, the reviewers should be considered so as to not push a bolus of documents for the same reviewers at the same time.

The project schedule can also be used to identify tasks that can be run in parallel, thereby shortening the timeline. This is especially useful when the content of sections is either completely independent or completely identical (allowing copy/paste across sections).

13.3.1.6 Lexicon

A lexicon can be a time-saver and help with consistency across the document, but I have had hit-or-miss success with it depending on the submission team. Some teams thought it would take too much time to create the lexicon to make it worth the ability to copy/paste. Others thought it was great and fully used the text blocks throughout the submission. It is really up to the medical writer's style. Within MS Word, there are ways to save text blocks in the Quick Parts function, and these can be used across sections when describing the same concept.

For example, the following concepts can be written once and saved to the Quick Parts library for re-use.

- General descriptions of product, program, and studies (note do not include data in these descriptions)
- About [DRUG]
- About [indication(s)]
- About [study1] – design and objectives
- About [study2] – design and objectives

13.3.1.7 Preliminary Key Messaging

The MAS should be a story of why the new drug is needed. It should describe the unmet medical need and the evidence that was gathered through the CDP and CMC Plan to satisfy that unmet need. In short, it is a cohesive narrative of why the regulatory reviewer should approve the new drug. To that end, key messages for product positioning and differentiation from available therapies should be woven into the story.

If the entire MAS was written by one or two authors, the storyline would be pretty easy to keep consistent throughout the document. However, in reality, there are dozens of authors, each coming from different disciplines and authoring capabilities. In addition, the messaging in the submission should align with the messaging in the supporting reports (e.g., CSRs). Therefore, it is helpful to align the submission team to a set of key messages early in the process. The format below may be helpful to create a key messages document.

Key message	*What is the product position or differentiating factor?*
Supporting data/evidence	*What evidence supports the message?*
Precedent and differentiation	*What do competitors have on their label, and how is your product different?*
CTD section(s)	*What sections will this message be applied to?*

Preparing the key messages document early in the process allows the team to identify the claims they want to make in the label and to ensure they will have the data to support the claim. Furthermore, even if the data are available, the submission team may need to perform additional analyses or create special tables to be able to present the data in the right way. This requires statistical programming that can take weeks to create, so it is important to get a head start.

13.3.2 After TLR

After a positive data readout of the pivotal trial(s), the pressure is on to complete the application as fast as possible. Some of the key activities besides authoring the last pieces to include the late-breaking data are as follows:

- Hold messaging meetings to align on key messages.
- Create the draft label text in the form of a US Package Insert or Summary of Product Characteristics (SmPC).
- Start working on the CSRs and patient narratives.
- Collect required information and documents such as investigator and IRB information and financial disclosure forms.

13.3.2.1 Final Messaging

After final data are available, the preliminary key messages should be revisited to make sure the data support the desired claims as originally defined. The key messages may need to be revised to accommodate the actual results, both favorably and unfavorably with respect to the target product profile. When revising the preliminary messages into final messages, it is important for the team to note areas where they anticipate the regulatory reviewer to challenge the claim, and to prepare responses to those questions in advance.

13.3.2.2 Draft Label

Creation of the key messages document goes hand in hand with the draft label text. The label is the ultimate output of the negotiation between the sponsor and the regulatory agency. It contains the claims that sales representatives can reference when discussing the product to consumers. Hence, it is important to have clear and definitive text in the label. Section 13.1.14.1 is where the draft label language is included in the NDA.

If a draft regulatory target product profile already exists, this can be a great starting point for the draft label. As discussed in Chapter 12, the Regulatory TPP contains all the headings of the US Package Insert, along with a section where annotations can be made about the evidence that supports the label language.

13.3.2.3 CSRs and Narratives

The CSRs and patient narratives for the pivotal study must be written in parallel with the NDA. The final key messaging that is used in the summary sections of the NDA should also be included in the CSR as much as possible. Often during the write-up of the CSR, there are new messages identified, such as when new analyses are run, and a differentiating factor is revealed. For this reason, the CSR is often left "open" while the NDA summaries are written, and both documents are finalized at the same time after all text has been reviewed and finalized.

Patient narratives are also an important piece of the CSR that is often get forgotten until the last minute. While many narratives can be generated programmatically these days, they still require careful review by qualified clinicians. The narratives are often references in the CSR, so the coordination of narrative writing is important for being able to complete the CSR.

13.3.2.4 Administrative Documents

There are several documents that are needed for the submission, such as financial disclosure forms, investigator information, and IRB information. The Clinical Operations team is usually responsible for collecting this information and entering it into a format that is usable for the CSR and submission. It is prudent to get an early start on these documents because sometimes investigators are slow to respond, especially for studies that were completed a long time ago, but inevitably, there always seems to be some documents and information that are left to the last minute.

13.4 SUMMARY

A marketing application represents the culmination of the work conducted under an asset's CDP and CMC Development Plan. The creation of a MAS requires tight coordination and collaboration between many cross-functional team members to align on storyline, messaging, and content of the document. While the best practices presented in this chapter were specific to a MAS, they can also be used for any other type of CTD submission such as an Investigational New Drug application in the United States to enable first-in-human trials or a Market Authorization Application in the EU or New Drug Submission in Canada to allow public sale of a drug in those regions.

Part 3

GRIDALL: A Comprehensive Framework for Managing Projects

14 Introduction to GRIDALL

Joseph P. Stalder
Groundswell Pharma Consulting

CONTENTS

A typical project management methodology (PMM) will facilitate knowledge management, repeatability, comparability, quality, and future impact. An organization that fails to develop and implement a PMM may jeopardize its productivity and potentially its overall success. I have seen several small- to medium-size biopharmaceutical companies and even some large companies, that have a few components of a PMM that arose organically over time, but they lack a comprehensive, planned methodology that enables end-to-end management of projects. I have also seen companies, especially large ones, where some project management components (e.g., risk management) are treated as stand-alone activities and thus missing the context of the broader project situation.

The next few chapters will provide a framework that can be used to create a comprehensive PMM for a biopharma company of any size. The key concepts to keep in mind are that a PMM needs to be *comprehensive*, meaning all major pieces of project information need to be included, and *connected*, meaning the pieces need to relate to each other so that the context is maintained.

14.1 INTRODUCING THE GRIDALL FRAMEWORK

GRIDALL is a project management framework that supports the development of a comprehensive and connected PMM. It can be used to build PMMs that sustain project management activities across a wide variety of industries and project types. It can be used to create a new PMM from scratch and to identify gaps in existing PMMs. It can also be used as a teaching tool for junior project managers and for advanced practitioners who are looking to improve their effectiveness.

DOI: 10.1201/9781003226857-17

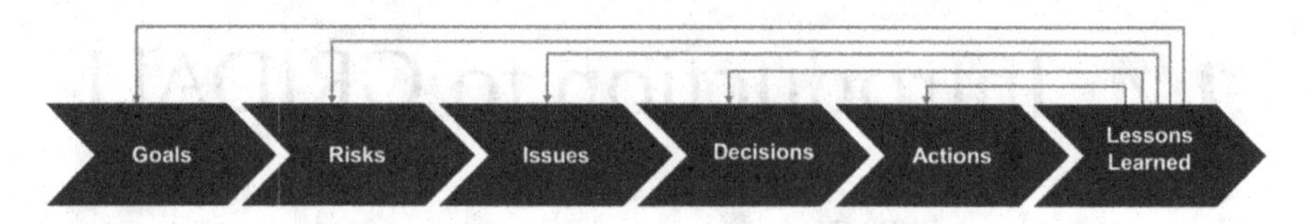

FIGURE 14.1 The GRIDALL Project Management Framework.

The GRIDALL model has six components: goals, risks, issues, decisions, actions, and lessons learned. Each component builds on the previous one in sequence; hence, the most valuable PMM will achieve all six components (Figure 14.1).

There are seven core tenets of the GRIDALL model:

- Every project has a goal.
- Every goal has risks.
- Every risk can become an issue.
- Every issue requires a decision.
- Every decision is followed by an action.
- Every action imparts an opportunity for lessons learned.
- Every lesson learned can be applied to future goals, risks, issues, decisions, and actions to improve repeatability, quality, and future impact.

While none of these components are novel in and of themselves, the GRIDALL model offers an improvement upon more common tools such as ADI logs and RAID logs. First, ADIs and RAIDs are not complete, leaving out some key project information (e.g., goals and lessons learned). Second, ADIs and RAIDs are not connected, and thus they do not provide a continuous flow of project information throughout the project management life cycle (e.g., risks are often considered in a vacuum, not in the context of goals; decisions are often captured as stand-alone statements, rather than in the context of the issue) – more on this discussed in the next section. Finally, ADIs and RAIDs do not provide a feedback loop, where lessons learned can be applied to future goals, risks, issues, decisions, and actions, and thus, they do not enable teams to continuously improve on their performance.

Fortunately, because the components are usually familiar to organizations and project teams, adoption of the GRIDALL frameworks generally requires a low level of change management effort. When rolling out the GRIDALL-based PMM, project managers will only need to explain the key concepts: comprehensive and connected. When I introduce GRIDALL to a project team, I like to describe it as "stuff we're already doing, just organized into a better flow".

14.2 GRIDALL AS A FRAMEWORK FOR BUILDING YOUR PROJECT MANAGEMENT METHODOLOGY

GRIDALL is a framework that allows PMOs to pick the tool that best meets the needs of their organization, and project managers to pick the best tool for the specific situation. As shown in Figure 14.1, there are many tools to choose from for each component. The set of tools that you use in your organization then becomes the PMM.

With smaller startup companies that have a clean slate, you may need only one tool per component to meet the needs of the organization. As you grow and the needs become more complex, you may need to add more tools. Take *Actions* as an example. A small company may need only a simple action tracker to ensure decisions are followed through appropriately. However, eventually, you may need to add a corrective and preventative action template when more formalized, inspection-ready documentation is needed.

After you have picked the tools for your PMM, I find it best to create a PM Playbook in slide form that has templates of each one. The PM Playbook is useful for many reasons:

- New project managers can quickly come up to speed with your company methodology.
- Existing project managers have a consistent starting point when they need a particular tool.
- The PMO can use the PM Playbook to educate others of the business who may want to know what PM tools are supported.

People sometimes ask if having too many tools in the PMM creates confusion. For example, is it confusing to have some groups using SMART goals and others using OKRs. In theory, yes, it is potentially confusing to have redundant tools. However, it is also important to realize the subtle differences between the potentially redundant tools and to enable your project managers to pick the one that best suits their situation. A clear example of this is in the *Risks* component, where risk analysis done by CMC PMs rely on a failure mode and effects analysis (FMEA), but for R&D PMs, a simpler risk register is more appropriate (Figure 14.2).

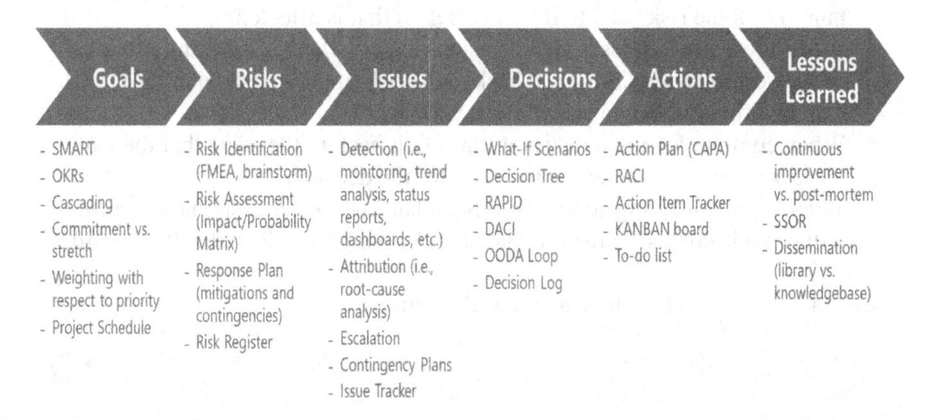

FIGURE 14.2 GRIDALL framework and potential tools to build a project management methodology.

14.3 GRIDALL AS A TOOL TO ASSESS YOUR ORGANIZATION'S PROJECT MANAGEMENT METHODOLOGY

The GRIDALL framework can be used to assess the completeness and effectiveness of your organization's PMM. When assessing your organization's PMM, consider the following:

- Do we have a clear goal-setting process? Are our goals shared transparently? Are our goals readily accessible?
- Do we manage risks effectively? Are risks documented (e.g., in a risk register) in a place that the team and other appropriate stakeholders can see?
- Do we identify and respond to issues effectively?
- Do we make decisions quickly and correctly? Are decision-making roles clearly defined and understood? Are decisions captured and shared quickly across the project team?
- Are decisions implemented effectively in the form of action plans with clear definitions of who, what, and when?
- Are we gathering lessons learned to share across the organization to improve future efforts?

If the answer to any of the above is "no", you have a gap in you PMM.

14.4 GRIDALL AS A PROJECT COMMUNICATIONS MODEL

Not only is GRIDALL useful for building and assessing your organization's PMM but it is also a very useful model for communicating project information. The key concept here is to share appropriate context via a technique I call "linking it back". In this way, the PM shares a new piece of project information by providing it in the context of the previous component. For example,

- When assessing and communicating risks to other stakeholders, it is important to link the risk back to the project goal that is affected.
- When presenting an issue to a decision-maker (whether it be an individual or a governing body), linking it back to the risk (if previously identified) will help by providing context to the issue.
- When communicating a decision, linking it back to the issue that the team faced will provide context to the need for the decision.
- When communicating actions, linking it back to the decision that was made will provide context to the rationale for choosing the selected path forward.

Figure 14.3 represents the "linking it back" technique.

FIGURE 14.3 Representation of the "linking it back" technique enabled by the GRIDALL Framework.

And in the final step, capturing and sharing the details of your journey through the GRIDALL process will enable others to learn from your experience. Lessons learned provide institutional memory that can serve you, your team, and other teams to be more effective when the same or similar situations arise again. Lessons learned provides a feedback loop to improve repeatability, quality, and future impact.

Using GRIDALL and the "link it back" technique enhances the PM's ability to communicate project information to team members, leaders, and stakeholders. As with any communication, context is key, and context can be easily provided by reviewing the associated topic from the upstream component.

The GRIDALL can also be used to help the team determine what information needs to be shared with what part of the organization. For example, if a risk is identified that affects a corporate goal, the risk should be shared with the executive-level committee. However, if a risk affects only sub-team goal, the risk can be managed within the sub-team. On my GRIDALL implementation, I have created a column titled "Flags" in the Risk and Issue components, and the entries may be Corporate, Core Team, CMC Sub-Team, Clinical Sub-Team, etc. I then have a report that pulls the appropriate risks (e.g., a portfolio report that contains only the corporate-level risks) for reporting.

14.5 HOW GRIDALL APPLIES TO BIOPHARMA PROJECTS

The GRIDALL framework can be applied to many types of biopharma projects, as follows:

- asset development projects,
- nonclinical development projects including pharmacology and toxicology studies,
- clinical development projects including clinical trials, and
- CMC development projects including drug substance, drug product, and analytical methods.

A special nuance of biopharma projects, as described in Chapter 16, is that risks can be categorized as "design" risks and "implementation" risks. The GRIDALL process is useful for implementation risks as it relates to the achievement of outputs, and not outcomes.

14.6 BEST PRACTICES FOR IMPLEMENTING GRIDALL IN YOUR ORGANIZATION

As mentioned previously, the GRIDALL framework is usually easy to implement because organizations are already using the components. The power of the GRIDALL framework is in the comprehensiveness, integration, and connection of the components. Therefore, when implementing GRIDALL in your organization, it is key to have a single tool that includes all components, starting with the Goals component and then populating downstream components in turn.

If starting from scratch, a simple spreadsheet (e.g., Excel) with one tab per component is easy to create and can be used effectively for a small number of projects. A PM can start with an existing ADI or RAID log, add the missing components, and add a few columns to create the linkages between components, and you have a good enough tool to apply the GRIDALL.

For a larger implementation, a solution built using a low-code/no-code application (e.g., Microsoft Power Apps) will be more effective. A solution like this is certainly more scalable and allows for better cross-project reporting in larger organizations. It also keeps the project information centralized in a single repository, which provides the following benefits:

- Data entry becomes more consistent.
- Reporting becomes easier.
- Institutional memory becomes easier to maintain.

It is important to keep in mind other systems that may integrate with the GRIDALL. A key integration is with the PMO's reporting system, which may require certain metadata in the GRIDALL to be able to generate the right content in the reports. For example, adding an option in the "Risk Status" that says "For escalation to governance" will be the flag for the system to pull that risk into the appropriate risk report.

As with many new tools, the best approach may be to start simple and build as you go. Remember that demonstrating value with the methodology is the key to successful adoption, and the tools only support the adoption.

14.7 SUMMARY

The GRIDALL framework provides a comprehensive and connected PMM that can be used across many types of biopharma projects. This framework provides an improvement over existing tools like ADI logs and RAID logs because it is more complete (i.e., contains goals and lessons learned), it connects between tools (e.g., a risk is associated with a goal), and it provides a feedback loop that promotes continuous improvement. GRIDALL can be used to assess an existing PMM, or to create a new PMM from scratch. It can be used as a model for communicating project information because useful context is available by linking one component back to the previous component. GRIDALL is easy to implement because none of the components are new to most team members; it is only the way the components are organized and linked that may require explaining to the team. GRIDALL implementation can start simple with a spreadsheet and grow into more robust and scalable tools like no-code/low-code solutions as an organization's maturity and sophistication allows.

15 Goals

Joseph P. Stalder
Groundswell Pharma Consulting

CONTENTS

15.1 WHAT ARE GOALS AND WHY SHOULD A PROJECT MANAGER CARE?

The purpose of a project is to achieve a goal. That goal could be the delivery of a product (e.g., an approved drug that is available for patients to use), a service (e.g., the ability to support the business with planning its resource capacity), or a result (e.g., dosing the first patient in a first-in-human trial). Setting goals is arguably the most important part of planning a project because it encourages the team to begin with the end in mind (see Habit 2 of Steven Covey's *The Seven Habits of Highly Effective People*). The project manager then takes this initial vision of the project outcome or output and breaks it down into actionable tasks. I like to think of the project schedule as a manifestation of the project's goals, and the project schedule is a clear piece of the project manager's toolkit for achieving project success.

In drug development projects, to achieve most goals will require multiple people working on concert – hence the formation of a project team. And wherever a project team is involved, project managers provide value by keeping all members of the

DOI: 10.1201/9781003226857-18

project team clearly focused on the goal. Project managers make project goals their own goals, and they use their skills and expertise to inspire a sense of shared purpose within the project team to ensure the goal is at the forefront of all decisions and actions.

In addition to goals being so important to projects and thus to project managers, project managers are also the most appropriate team member to drive of the goal-setting process. Here are a few reasons why:

- Project managers have a holistic view of the project (i.e., all activities, all milestones, and all outputs).
- Project managers are not hindered by functional constraints or siloed departmental thinking.
- Project managers are experts at creating and driving a process.
- Project managers are the central hub of all project information.

And so, with goals being the most important part of planning a project and project managers being best suited to driving the goal-setting process, an important role the project manager has to perform appropriately is the goal-setting duty. The table describes some basic tenets of goal setting, and the corresponding role that a project manager plays in addressing that tenet.

Basic Tenet of Goal Setting	Project Manager's Role
Goals should define the company's priorities	PMs need to ensure the project teams have the same relative priorities for their goals
Companies should use a consistent goal-setting framework	PMs need to know the framework in order to drive the process for their project team
Goals should cascade from top levels of the organization all the way to the individual contributor level	PMs need to ensure project team goals reflect the corporate goals and that team members understand their part in achieving the goal
Goals should align groups of people to achieving a common result	PMs need to coordinate cross-functional planning (i.e., interdependencies) and delivery of team goals
Goals should be clear and transparent	PMs need to be the champion and keeper of project team goals

This chapter begins by describing some of the history and theory of how goals apply to today's business world. We will then look at a practical process for setting project team goals that meet the basic tenets described earlier. We then deep dive into a biopharma-specific goal concept – product requirements as defined by the target product profile (TPP). And finally, we will evaluate some pros and cons of the typical annual goal-setting process and how to navigate the challenges in today's business culture.

Note that we are focusing on project goals, and not organizational goals such as the company's financial position or environmental sustainability efforts. This is because project managers have the most influence and are concerned most with project goals.

This would be rare, but if you are ever at a company that does not have transparent, documented goals, you, as the project manager, need to push for having them. The following principles can help your argument for establishing a goal-setting process at the company:

- Project goals set the foundation for all work within the organization.
- Project goals reflect a company's priorities and focus the company's energy on what needs to be accomplished.
- Prioritized project goals allow teams to make better decisions when trade-offs are needed.
- Project goals are essential to good project management.

15.2 SOME HISTORY AND THEORY OF GOAL-DRIVEN ACTIVITIES

It is hard to imagine any enterprise not having goals. I am sure even ancient Egyptian rulers had a goal of delivering a pyramid by a certain time point. But today's business is driven by different motivational concepts than ancient times. And indeed, even within the past few generations, we hear of changes to the motivational construct for achieving work (e.g., millennials being less motivated by pay and more by a psychological connection to their work). Nonetheless, today's businesses were shaped by a few major themes and concepts in the last half of the 20th century. We will highlight a few of the publications that drove the goal-setting language of our time and then look at some of the concepts that apply to most businesses today.

15.2.1 Publications on Goal Theory That Every Project Manager Should Know About

There are a few key books that set the standard for how goals are talked about in today's business world. While certainly not an exhaustive overview, I think of these publications as being the most influential in the sense that most biopharma companies use at least some aspects of the concepts described. As drivers of the process, it is important for project managers to know the foundations of today's goal-setting principles.

We start in 1954 with Peter Drucker's book *The Practice of Management*, in which he describes the concept of management by objectives (MBO). Drucker aimed to increase company performance by aligning corporate goals with team and individual goals. Specifically, high-level, strategic goals set by senior leaders should be broken down into more discrete and actionable objectives that are owned by project teams or individual employees, and on which their performance will be evaluated. This concept of breaking down high-level goals is what I later describe as *cascading*. The MBO approach has dominated the business landscape for several decades, but it has recently been challenged as being too focused on delivery of outputs and not enough on innovative thinking. It has also been criticized as being too inflexible, leading to "zombie" projects that have little value but are being worked on merely to satisfy a goal.

In 1968, Edwin Lock published a paper called *Toward a Theory of Task Motivation and Incentives*, which he and Gary Latham later turned into a book called *A Theory of Goal Setting and Task Performance* (1984). The book popularized the concept that clear and challenging goals, along with appropriate feedback, contribute to higher and better task performance because employees understand what needs to be done and why their actions contribute to a broader purpose. This may be considered the origin of the SMART framework that is common today, but credit goes to George T. Doran in 1981 for coining the acronym SMART in his article titled *There's a S.M.A.R.T. way to write management's goals and objectives.* We will dig deeper into the SMART framework later.

In the 1994 book *Built to Last: Successful Habits of Visionary Companies*, Jim Collins and Jerry Porras describe a characteristic that is common to successful companies: big hairy audacious goals (BHAGs). In their words, "All companies have goals. But there is a difference between merely having a goal and becoming committed to a huge, daunting challenge…". BHAGs are goals that define a vision for the future and drive progress forward by serving as a unifying focal point of effort. BHAGs have a time horizon of 10 or 20 years, and drawing on President John F. Kennedy's 1961 goal to land a man on the moon by the end of the decade, they are commonly referred to as "moonshots". I'll describe moonshots more below.

Fast forward to 2018 with John Doerr's book on the OKR goal-setting framework *Measure What Matters.* In it, Doerr describes the concept of OKRs and their real-life application to companies like Google and Intel. He then goes on to describe the OKR framework and four superpowers that make the OKR method successful (i.e., focus and commit to priorities, align and connect for teamwork, track for accountability, and stretch for amazing). Interestingly, contrary to Drucker's MBO theory, Doerr proposes that individual goals should not be linked to the performance merit system because this process limits creativity and bold approaches, rewarding instead the achievement of status quo. Rather, individuals and managers should engage in continuous performance management through conversations, feedback, and recognition.

Taken together, these publications set the foundation of key goal-setting principles that are present in most of today's biopharma companies: cascading, SMART, moonshots, and OKRs. We will describe the practical use of these concepts in the sections below.

15.2.2 What Is the Difference between Goals, Objectives, and Requirements?

When discussing goals, I like to distinguish the terms goals, objectives, and requirements. However, admittedly, it is difficult to persuade the general community to adopt these definitions, and I end up using the language that is familiar to most stakeholders (i.e., they are all "goals").

- **Goal**: Goal is a desired outcome that you want to achieve. Goals are visionary, and they are described in vague or lofty terms. An example of a goal is the purpose statement in a clinical trial protocol that describes why a study is being conducted (e.g., to compare an investigational drug X with standard-of-care drug Y in disease Z).

- **Objective**: Objective is a specific output that you plan to deliver. Objectives are practical, and they are described in discrete (i.e., SMART) terms. This is akin to the objectives of a protocol, where every objective should have a measurable endpoint that is used to judge the success of the clinical trial (e.g., To demonstrate a meaningful improvement in overall survival for drug X compared with drug Y at 24 months).
- **Requirement**: Requirement is a feature, function, technical aspect, or characteristic that must be satisfied by completion of the project. Requirements can also be referred to as *specifications*. The document that captures requirements in drug development is the TPP. More details on TPPs are given later.

15.2.3 What Are Cascading and Aligning, and Why Are Those Important?

The concepts of cascading and aligning are meant to keep everyone in a company focused on the important work that needs to be done. The idea is that the business verticals need to be working on the same goals, and the horizontals need to be aligned with each other on who does what and when. While Drucker's work on MBOs calls all of this "aligning", I like to distinguish the term *cascading* to apply to the business verticals and *aligning* to the project team horizontals.

As described in the MBO concept that Drucker wrote about in 1954, corporate strategies are created at the highest level in the organization and handed down to smaller groups who then break down the strategies into actionable elements (e.g., goals and action items). At the project team level, goals may be cascaded from corporate, business unit, or divisional levels, depending on the size and complexity of the organization. Regardless of where it came from, the project team level is commonly where *goals* are turned into *objectives*.

Within a group (e.g., project team or department), goals serve to align members to achieve a common result. Goal alignment within a project team is especially important due to the underlying tension between project team membership and functional group membership. When individuals have misaligned project and functional goals, it is often the project goal that gets deprioritized when demand for functional goals forces a trade-off decision. This is because performance management is ultimately cast by the individual's functional manager, and not the project manager.

15.2.4 What Are Committed, Stretch, and Moonshot Goals?

No all goals are alike, so let us define a few terms that are commonly used to distinguish to context of a given goal. When discussing these goals throughout a project team or an organization, it is useful to have a consistent terminology so that the goal's essence and time horizon are understood. Many people have heard about "committed", "stretch", and "moonshot" (or BHAG (Big, Hairy, Audacious Goals)) goals; however, my definitions may differ slightly; as they relate to the drug development project teams, I want to spend some time defining the terms.

Committed goals are those in which the organization has allocated the resources needed to meet the goal. In practice, this means that the company's governing bodies agree that the goal must be achieved for the company to be successful, and they are willing to allocate a set amount of the company's resources and energy to meet the goal. Committed goals should be achievable, and failure to meet the goal requires explanation for the miss as it shows errors in planning or execution.

The conventional definition of a stretch goal is that it is something that is very ambitious and less likely to be achieved. However, I believe all goals should be and are achievable, given the right resources are allocated to the project. Thus, in my definition, the difference between "committed" and "stretch" lies in the resources and energy the company is willing to allocate, and not on the level of risk associated with the goal. Stretch goals then describe a situation where if additional resources were applied, upside opportunity may be achieved. The stretch/upside component can be in terms of reduced time (e.g., achieving first patient dosed milestone 1 quarter earlier), greater yield (e.g., producing larger volume of drug substance through additional manufacturing runs), or higher quality (e.g., providing a better clinical trial dataset by performing additional cleaning activities). In each of these situations, more resources, whether that is human capital or financial capital, are required to produce the upside.

How stretch goals can play out when setting goals is that project teams can propose a stretch goal to leadership to demonstrate the team's most ambitious plan and the resources required to achieve that goal. This gives leaders in the organization an option to improve on the delivery of a goal if they are willing to allocate additional resources. For example, a company may have a committed goal of dosing the first patient in a pivotal clinical trial by Q2. However, if an additional resource was applied, say a medical writer to expedite protocol development, or support from a study startup unit to activate sites quickly, the goal may be achieved in Q1. Leaders in the organization can then weigh the option of providing more resources to achieve this 1-quarter time savings.

The term "moonshot" is often used to describe a goal with a time horizon of a decade or more. For many small biopharmaceutical companies with a cash runway of less than 2 years, it is hard to imagine a future beyond 3–5 years. However, for larger companies that have resources to apply to multiple "shots on goal", these moonshots can provide a longer term vision for treating a certain disease that supplants a project with a shorter term turnout. For example, a company could be working on a small molecule or biologic therapy to reduce the symptoms of an autoimmune illness and at the same time be working on a gene or cell therapy intended to cure the disease. The symptom reduction approach could be achieved in 3 to 4 years, whereas the curative approach may still be 10–20 years away.

15.2.5 Why Should We Use a Goal-Setting Framework, and What Options Do We Have?

Having a consistent goal-setting framework leads to better organizational performance because it ensures that all levels of the organization set goals in the same way and cascade and align goals throughout project teams and across functional lines.

In addition, a clear and transparent goal-setting process promotes engagement and support from the teams and individuals who will end up doing the work.

There are many goal-setting frameworks, such as:

- SMART
- OKR
- Do Now/Do Soon/Do Later
- MoSCoW (must do, should do, could do, won't do)

Of these, SMART goals are probably the most common, and this framework works well for setting project goals in biopharma companies. There are many variations of the definition of SMART, and one, perhaps the most widely used, is below:

S	**Specific**	The goal clearly defines the outcomes to be delivered, with any necessary interpretation agreed upon by the team and leadership.
M	**Measurable**	The achievement of the goal can be objectively assessed according to pre-determined and applicable measurement.
A	**Attainable**	The team has the resources, time, circle of influence, and information needed to allow it to achieve the goal.
R	**Relevant**	The goal addresses work and results that clearly align with the goals of the next-higher level of the organization.
T	**Time-bound**	The goal clearly specifies the delivery date or schedule.

OKRs are relatively newer to the life science industry but deserve mention here due to increasing popularity and success shown in the technology sector. The origin of OKRs can be traced to Peter Drucker's 1954 theory of Management by Objectives, but it was Intel co-founder Andrew Grove in the 1980s who fine-tuned the MBO concept to what we now know as OKRs. At Intel, Grove utilized an adapted MBO concept where each objective is further broken down to a few key results that are used to measure progress and achievement.

In the late 1990s, John Doerr, who observed the OKR framework in practice at Intel, convinced Google to use the OKR framework, which then provided evidence of its value at a scalable, innovative, fast-paced environment. Google later produced a video and playbook that describes their approach to OKRs, igniting rapid uptake of OKRs throughout the tech industry. Doerr would go on to publish the framework in his 2018 book "Measure What Matters". The adoption of OKRs in the life sciences industry seems to have begun more recently, perhaps stemmed by the COVID pandemic and resulting change in the nature of office-based vs. remote workforce. Even still, I commonly see corporate goals being driven by the SMART framework, and only pockets of the organization (departments or project teams) that use the OKR framework.

The OKR model focuses on what do you want to achieve (objective) and how are you going to do it (result). The objective is a high-level statement of the desired outcome, and the results are the measurable benchmarks you will use to track the path to achieving the objective. For us in biopharma, it is easy to think of them as the objectives and endpoints that we see in protocols. Typically, there are three to five results per objective.

Objective	Key Result
What do you want to achieve?	How will you know you are achieving it?

A key premise behind OKRs is that as Doerr puts it, "Transparency seeds collaboration". In theory, making goals public across the workplace enables others to know what a team or individual is working toward, which then promotes further cascading and alignment. Furthermore, transparency of goals across the organization also creates a shared sense of purpose where teams and individuals can clearly see how their work contributes to a broader corporate strategy.

15.3 A THREE-STEP PROCESS FOR SETTING PROJECT GOALS

Most companies employ an annual goal-setting process, where goals are set at the beginning of the year to guide the company's efforts throughout that year. I will discuss pros and cons of this yearly outlook at the end of this chapter, but for purposes of practicality, let us assume the definitions and process described below are in the context of this "year-at-a-time" horizon.

The goal-setting process can be broken down into three steps (Figure 15.1).

The project manager should be the primary driver for each of these steps, and the descriptions below will give you a practical guide for managing the goal-setting process through these three steps.

15.3.1 Identify

In the annual goal-setting process, goal identification often begins 1–3 months before the goal-year starts. When developing project goals, development teams should consider the following sources:

- corporate goals set by senior management;
- development plans set by project team, particularly focusing on key decision points in the development plan such as first subject dosed, data readout, or regulatory submissions;
- intermediate deliverables set by functions within the project team, such as clinical pharmacology data becoming available that will affect eligibility criteria in the clinical protocols.

FIGURE 15.1 Three-step process for setting project goals.

By considering these three sources, the project team is able identify goals in a "top-down" and "bottom-up" approach. After an initial draft of the goals is ready, the project leader and project manager should propose them to senior management to get feedback or agreement and then take that information back to the project team for discussion. There may be a few iterations like this until the project team goals and the corporate goals are set.

I like to call this approach the "top-down, bottom-up, meet-in-the-middle" approach. Through this approach, the expectations of major stakeholders will be incorporated into the goal-setting process, and project teams and functions will be allowed to give input into what they feel is achievable.

15.3.2 Refine and Align

After high-level goals have been identified, the project team can then refine and align the goals. This is typically done 1 month before the end of the previous goal-year and can continue into the first few weeks of the goal-year.

Questions to consider when refining goals:

- What is the actual deliverable we are trying to produce?
- Who will "own" the goal to act on it and monitor its progress?
- How does the goal align with broader corporate goals?
- What are the assumptions, risks, and voiding criteria for the goal?

All functions and sub-teams within a project team should have the opportunity to pressure-test the goal to make sure it is feasible. Functional lines contribute planning information such as estimates of time, cost, resources required, assumptions, and risks. For example, the goal of first site activated may be challenged by the CMC sub-team because they may not be able to get clinical trial materials to sites in time. The purpose of aligning across the project team is to reach consensus on goals as needed to ensure all sub-teams/functions agree the goal is doable.

The following template is useful for refining and aligning project team goals.

Goal title	What is the desired outcome?
Deliverable	What is the result or output?
Delivery date	When will the result or output be delivered?
Stretch	If given more resources, what upside could be achieved?
Owner	Who will act on and monitor progress toward the goal?
Strategic alignment	How does the goal associate with the broader corporate goals?
Assumptions	What has the team assumed will happen or be in place to achieve the goal?
Voiding criteria	What circumstances would cause the team to have to redefine the goal?

Here are a few things to note on the goal details table:

- Stretch: Not all goals will have a stretch opportunity. For example, a 26-week chronic toxicology study will take the same amount of time, regardless of adding more resources to it.
- Strategic alignment: This is useful for discussions with senior management on how the goal fits in with the corporate goals. The concept of cascading is documented here.
- Assumptions: These are very important for conveying things that are out of the team's control that must occur before the proposed goal can be met. As with all future-looking activities, there is a chance that one or more of the assumptions may not come true, which then becomes a risk we will want to monitor.
- Voiding criteria: These are useful for keeping goals nimble and flexible based on circumstances beyond the project team's control. For example, if a pre-clinical program with a goal of first patient dosed in Q4 demonstrates a major toxicity that should terminate the program, the goal should be voided so that the team does not feel like they need to keep working on it just to satisfy the goal. Voiding criteria should not compensate for inadequate planning and execution. Instead, they should be used to refocus a team if scientific, regulatory, or business reasons make the goal not worth pursuing any longer.

After all goals are agreed and plans are confirmed across all functions and sub-teams in the project team, it is time to lock them down and start communicating them out.

15.3.3 Finalize and Socialize

Goal finalization usually occurs in the first few weeks of the goal-year when the board of directors agrees to the corporate goals, thus setting the stage for project team goals to also be locked in. At the project team, the project leader and project manager will present the final goals to the rest of the team at a core team or project team meeting. Confirmation of the final goals sets off a responsibility of your project team members to socialize the goals to functional line leads and to the extended project team.

- Functional representatives on your project team should communicate the final goals to their line leads so that budgets can be set accordingly.
- Sub-team leaders and functional representatives can start to communicate the final goals to the extended team to ensure plans are put in place to deliver on the goal.

This socializing step is very important to make sure everyone involved in your project is on board with the goals. Although everyone on the extended team may not have been involved in creating the goal, successful execution requires all of them to be engaged and supportive of the goal, so making them not only aware of the goal but also aware of the need for the goal's output or outcome will galvanize the broader team members to deliver.

15.3.4 Next Steps for the Project Manager

After goals are finalized, the project manager should memorialize the goals in a format that allows for tracking progress and monitoring risks. While a few digital tools are available to support goal tracking, most companies still use manual processes (e.g., a table of goals on a slide). I find the tabular format below to be most useful because of its simplicity and clarity. Using this table on a slide, I check in with the project team every quarter and update the status for each goal. Usually, there are no surprises, but I have had situations when the check-ins revealed risks to a goal that needed intervention.

Goal	Owner	Due Date	Status

A few notes on the goal tracking table above:

- Goal: This should be a simple statement that combines the goal title and the deliverable from the goal details table above.
- Owner: This should be an individual, and not a group. For example, a sub-team leader or a functional representative.
- Due date: This should be the same as the delivery date from the goal details table.
- Status: The following status definitions are useful:
  - On track = blue
  - At risk = amber
  - Achieved early = dark green
  - Achieved = green
  - Achieved late = light green
  - Missed = red

As you go through the year and goals are achieved, it is important to recognize and celebrate the accomplishments. Again, the project manager is best positioned to plan the celebrations. If your company does not have specific project budgets to accommodate rewards, swag, and parties, ask your PM department head to add some budget to the department overhead.

15.4 PROS AND CONS OF ANNUAL GOALS

Most companies set annual goals at the beginning of the year, and teams strive to achieve those goals throughout the year. The "year at a time" goal-setting process works well from a corporate perspective because the timing aligns with the annual operating plan (e.g., annual budget), corporate reporting cycles (in the form of SEC filings), and performance reviews and bonus payouts. However, in drug development, where projects typically last several years, there may not be a major milestone that warrants a goal in the current year. Therefore, project teams need to consider a longer view on the outcomes and outputs that are expected, typically 3–5 years.

One of the problems with an annual goal-setting process is that goals are defined at the beginning of the year, and there is a perception that they cannot be changed. While committing to a goal is a noble concept and worth attempting, there may be times when a goal needs to change, and a company's goal-setting process should be flexible enough to accept revisions mid-year. In addition, an inflexible goal-setting process can result in project teams working on "zombie projects", ones that should have been terminated but the team feels obligated to keep working on it to satisfy the goal. Hence, when setting annual goals, there should be an option to void the goal so that the organization's energy can be refocused on more valuable efforts.

Another common problem with the annual goal-setting process is the tendency for senior leadership to squeeze a goal into a year, which puts extra pressure on the project team. For example, the project team works up a plan to deliver study results in Q1 of the next year. However, senior leadership accelerates the goal into the Q4 of the current year because the positive news flow will look good for the company. That one-quarter difference however can put the goal at risk of not being delivered, or being delivered with poor quality, or burning out the project team. Impossible goals end up demoralizing and demotivating the team, setting the stage for a bad company culture.

15.5 SUMMARY

All projects should have clearly defined and agreed-upon goals. Goals serve to align groups of people on what needs to be done, between senior leadership and project teams, between functional representatives on a project team, and between functional area heads and project managers and project leaders. There are many ways to set goals; the important thing is to have a consistent definition and process for creating goals. Project managers can use a simple three-step process of identifying, refining and aligning, and finalizing and socializing goals to set goals with their project teams.

16 Risks

Joseph P. Stalder
Groundswell Pharma Consulting

CONTENTS

16.1 INTRODUCTION

Every goal has its risks. PMI's PMBOK (7/e) defines a **risk** as "an uncertain event or condition that, if it occurs, has a positive or negative effect on one or more project objectives". Risk is inherent in the forward-looking nature of goals, and this is especially true in the complex and unpredictable world of biological systems and pharmaceutical development.

In drug development, I like to distinguish two types of risks: **design risks** and **implementation risks**. Design risks are inherent to the experimental nature of drug development, and they relate to the unpredictable *outcomes* of these experiments (e.g., the success of a pivotal clinical trial). Implementation risks, on the other hand, relate to the *outputs* of a team's efforts (e.g., the delivery of a top-line report to enable a go/no-go decision for submission), and they can be managed through a risk management process.

As we saw in Chapter 12, it is important and useful for project teams to quantify design risks in order to risk-adjust enterprise-level budget and resource planning and to support risk/return analysis in portfolio-level decision-making. However, the GRIDALL really is built to handle implementation risks that threaten the project team's ability to deliver an output.

Given a project manager's primary role to deliver successful projects and the fact that risks are things that could disrupt a project plan, project managers should be

DOI: 10.1201/9781003226857-19

concerned about and equipped to manage implementation risks. Not only that, but project managers, with their overarching view of a project's functional workstreams, are in the best position on the project team to identify disconnects in the project plan that can threaten project goals. Thus, the project manager is the most appropriate team member to create and facilitate of the risk management process. In this chapter, we will describe the common types of risks in biopharma companies and with biopharma projects and then outline a process for managing implementation risks. We will finish with a description of some best practices for keeping the risk management process finely tuned to be useful and convenient for the project team.

16.1.1 General Risks of Doing Business in Drug Development

Some risks are broad and generalizable to the drug development industry as a whole, and some are risks that are specific to an asset, indication, or clinical trial. I call the former "risks to doing business" in drug development. We can see some examples of the broad risks by looking at the risk factors section a typical SEC filing from a biopharmaceutical company. Some examples are below:

- All of our product candidates are subject to extensive regulation, which can be costly and time-consuming, cause delays, or prevent approval of such product candidates for commercialization.
- Preclinical development is uncertain. Our preclinical programs may experience delays or may never advance to clinical trials, which would adversely affect our ability to obtain regulatory approvals or commercialize our product candidates on a timely basis or at all, which would have an adverse effect on our business.
- The results of preclinical studies and early-stage clinical trials may not be predictive of future results.
- Our product candidates are in development. As a result, we are unable to predict whether or when we will successfully develop or commercialize our product candidates.
- Our product candidates may cause undesirable side effects or have other properties that could delay or prevent their regulatory approval, limit the commercial profile of an approved product label, or result in significant negative consequences following marketing approval, if any.
- The successful commercialization of our product candidates, if approved, will depend on achieving market acceptance, and we may not be able to gain sufficient acceptance to generate significant revenue.
- We rely upon third-party contractors and service providers for the execution of some aspects of our development programs. Failure of these collaborators to provide services of a suitable quality and within acceptable timeframes may cause the delay or failure of our development programs.
- Competition in our targeted market area is intense, and this field is characterized by rapid technological change. Therefore, developments by competitors may substantially alter the predicted market or render our product candidates uncompetitive.

- We are subject to competition for our skilled personnel and may experience challenges in identifying and retaining key personnel that could impair our ability to conduct our operations effectively.

While a project manager should definitely be aware of these "risks of doing business", this type of risk is often outside the control of the project team. They are for the C-suite of the company to worry about. Therefore, we will not focus on these broad risks in this chapter. Instead, we will focus on risks to assets, indications, and clinical trials because these are risks that the project manager and the project team can influence.

16.1.2 The Unmanageable "Design" Risks

Like the "risks of doing business" that are inherent in the biopharma industry, there are risks inherent to developing an asset into a commercially available treatment. These inherent risks can be categorized into three groups:

- risks associated with the design of the clinical trial,
- risks associated with the design of the data package that will be used for health authorities to form an opinion on market authorization, and
- risks associated with the design of the drug being developed.

I consider these risks to be "unmanageable" because they involve designs that are "locked" and cannot be changed. A clinical trial's design is locked by the protocol. A submission data package is locked by the clinical development plan. A drug product's design is locked by quality and compliance specifications. Therefore, the level of risk associated with these elements is related to the confidence in the assumptions used when designing the element. There is no way to completely remove these risks through mitigation tactics, so instead we adjust the project's valuation in the long-range plan (LRP) using a risk estimation called PTRS. This is described more in Chapters 5 12.

16.1.3 The Manageable "Implementation" Risks

While design risks are unmanageable because the parameters and specifications are locked by design, implementation risks are manageable in the sense that their impact can be reduced or avoided through proper risk management processes. Implementation risks are those associated with how things are done, and generally, there are a number of ways that things can be done that the team can choose from throughout the drug development life cycle.

16.2 A FOUR-STEP PROCESS FOR MANAGING RISKS

While project managers are best suited for running the risk management process, successful risk management needs to involve a broad community of stakeholders, including team members, sponsors, functional area heads, and executive leadership.

FIGURE 16.1 Four-step process for managing risks.

To align the understanding and expectations of these stakeholders, the project manager should create and socialize a risk management plan (RMP) that describes the process for managing risks.

The RMP should describe the process for identify risks, assessing the risk's severity, addressing the risk with a response appropriate to its severity, and then monitoring the response until resolution. As a recommended practice for those introducing risk management to a company or team, start simple so that you don't overwhelm your stakeholders with complexity and then build over time as the business need arises (Figure 16.1).

The sections below will walk through each step of this process. Using the concepts outlined here, you will then be able to create an RMP for your project team.

16.2.1 Identify the Risk

Risk identification is the cornerstone of the risk management process, but it is also the step that faces the greatest challenge. As mentioned previously, all team members and stakeholders should look out for risks, but a common challenge project managers face is that some team members and stakeholders may feel uncomfortable raising their concerns when they do identify a risk. There is an emotional element to raising a risk because they worry that their concern may seem negative or not supportive of the plan. When your team members don't feel comfortable sharing what could go wrong with a plan, the entire risk management process is undermined.

To overcome this challenge, it is important for the project manager and project leader to create a culture of transparency where risks are openly discussed. One way we project managers can engender this culture is to normalize risk identification by posing the question "Are there any risks to the plan?". Simply asking the question may prompt a team member to voice a concern.

While all team members and stakeholders should be involved in risk identification, project managers and project leaders are at a particular advantage for identifying risks because they have a cross-functional perspective and holistic view of the project and because they may have access to lessons learned from other project managers who have gone through the similar situations. Below are some sources the project manager and project leader can consider when identifying risks.

- Review the Project Plan
  - Look at critical path, then near-critical, and finally non-critical path tasks
  - Scrutinize tasks with many dependencies
  - Consider never-been-done-before activities and highly complex activities
  - Evaluate aggressive cost and duration estimates

- Assess Your Team
    - Consider your team's experience and expertise
    - Be aware of your team's capacity and conflicting commitments
- Brainstorm with Experts
    - Meet with key stakeholders, such as executive sponsors, senior leaders, and functional area heads, to uncover risks
    - Have other project managers review your plan
    - Check in with subject matter experts who have gone through the process before and identified lessons learned and best practices

Once a risk is identified, the project manager should start to record information about it in a risk register. The risk register is a comprehensive risk tracking log that enables the project manager and team to view all project risks in a central location. This allows not only a place to reference all known risks to the project but also to compare risks and prioritize the ones that should be actively addressed.

When describing risks in the risk register and in project status reports, I find it helpful to relate the risk to a project goal. This not only provides the context of the risk but also makes it easier to determine which risks should be communicated to the right stakeholders. For example, if a risk threatens a corporate goal, that risk should be included on a report that goes to the company's executive committee, whereas, if the risk is to a sub-team goal, it should be reported only to the sub-team.

Risk description	Affected goal	Impact	Probability	Exposure	Response	Owner	Trigger date
If <event> occurs, then <impact`>, resulting in <effect on goal>.	*The goal that will be affected if risk is realized.*	*1 = Low 2 = Medium 3 = High*	*1 = Low 2 = Medium 3 = High*	*Impact x Probability (see risk matrix)*	*Avoid, Accept, Mitigate, Transfer*	*Person responsible for monitoring risk and communicating risk status*	*Anticipated timing that risk could occur*

16.2.2 Assess the Severity

Once a risk is identified, the team should assess its severity or the degree to which the risk exposes your goal to failure. It is important to use a consistent methodology and framework with your team so that they become familiar with the process, and so risk assessments are comparable across projects. There are numerous methodologies for assessing severity, and all methodologies center on two dimensions: probability and impact.

When assigning probability and impact scores, I have seen various scales from 1 to 3, 1 to 5, and even 1 to 10. The argument for a larger scale is that it is then easier to tease out which risks need to be prioritized. However, in my experience, I have found that a simple 1–3 scale (low–medium–high) scoring assignment works best for drug development projects because it aligns nicely with how to address the risk: monitor, plan, and respond.

16.2.2.1 Scoring the Probability

The probability assessment will depend on your company's risk appetite. Small companies with little room for error are typically more risk-averse, so the thresholds are set lower. Larger companies with a larger portfolio of opportunities are typically more risk-tolerant, so the thresholds are set higher. Sample definitions for low, medium, and high are provided below.

Probability score	Risk-averse company (%)	Risk-tolerant company (%)
1 – low	<20	<33
2 – medium	20–50	33–66
3 – high	>50	>66

16.2.2.2 Scoring the Impact

The impact assessment can be more complicated than scoring the probability. I have seen some risk registers that specify the impact on various parameters:

- Time
- Cost
- Quality (e.g., some key factor of the TPP)
- Strategic value
- Peak sales
- Corporate image

To me, the most important consideration is the risk to achieving the goal using the approved plan. For example, as we saw in Chapter 11, when a governing body approves a clinical trial, the project manager can capture the key assumptions in a Governance Summary that can then be used as the baseline for comparing variances and risk assessments.

Like the probability scores, it is useful to have a consistent set of thresholds for the impact. Below are examples of thresholds:

Impact score	Time	Cost (increase vs. approved budget)	Quality
1 – low	< 1 month	<2% or <$100,000	Threat to minor factor of the TPP
2 – medium	1–3 months	2%–5% or $100,000 to $250,000	Threat to major factor of the TPP
3 – high	> 3 months	> 5% or > $250,000	Thread to many major factors of the TPP or to a key differentiation factor

16.2.3 Form a Response

After a risk has been assessed for probability and impact, the next step is to form a response strategy. A helpful tool for starting this process is the probability–impact matrix. Using this tool, the impact and probability are plotted on a grid, and the resulting location on the grid will guide the team on the type of response to form: monitor, plan, or respond.

- **Monitor**: No immediate action is needed for these risks. The risk owner will notify the team if the risk realizes or if the severity changes.
- **Plan**: No immediate action is needed for these risks, but a contingency plan is put in place to respond to the risk if it does realize.
- **Respond**: Immediate action is needed for these risks, in the form of both mitigation plans and contingency plans.

The probability–impact matrix can also be used to assess the project's overall risk profile. When all risks are plotted on the grid, the team can see how many risks are in the red zone, yellow zone, and green zone, which in turn can give you an idea of your overall project risk (Figure 16.2).

During this step, it is also important to identify the indicators or triggers for each risk. The risk owner should consider the following:

- How will you know the risk has occurred or is about to occur?
- When or how often should the indicator be monitored?
- If not the risk owner him/herself, who should monitor the indicator?

For high-exposure risks that warrant an immediate action, possible response strategies include the following:

- **Accept**: monitor the risk and form a contingency plan to execute in case the risk does occur
- **Avoid**: eliminate the risk either by not pursuing the goal or choosing another path to reach the goal
- **Escalate**: shift ownership of the risk to a higher level of the organization where it is more effectively managed
- **Mitigate**: reduce likelihood of risk occurring or the impact on the goal if the risk does occur
- **Transfer**: shift ownership of the risk to another party who will manage the risk and bear the impact if the risk does occur.

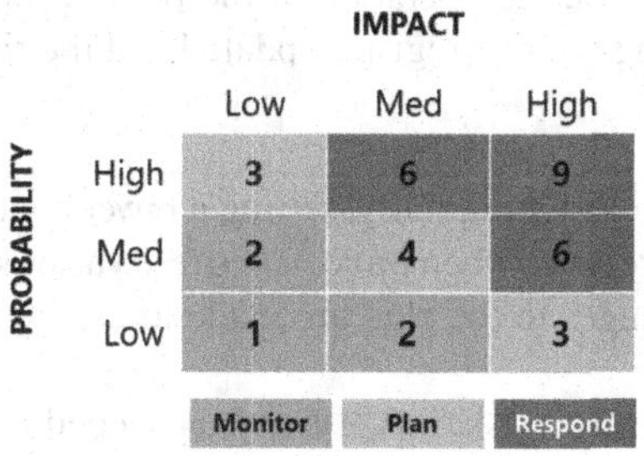

	Low	Med	High
High	3	6	9
Med	2	4	6
Low	1	2	3

FIGURE 16.2 Example Risk Probability-Impact Matrix using 3-level risk categorizations (low, medium, high).

An important concept with risk action plans is that not all risks warrant a mitigation. I see so many risk registers that include "risk mitigation" as the header, but the actual response is not a mitigation. To me, the more appropriate header is "risk response", of which one possible response is to mitigate the risk. An equally useful response is to form a contingency, which is an action plan that can be deployed by the team if the risk occurs (i.e., becomes an *issue*). This plan should be pre-approved by a governing body to allow the team to use resources and budget to respond to the issue without needing another governance approval. In some companies, these contingency plans are called "buy-ups", meaning it is part of the management reserve and contingent on the trigger occurring, but not included in the annual budget and resource plan.

16.2.4 Monitor Until Resolution

After a response is formed, the project manager should hold the risk owner accountable for monitoring the risk until it is fully resolved. There may be several ways for the project manager to do this, but I find 2 approaches to be sufficient: (a) include risks in project status reports to maintain visibility and (b) discuss risks with the risk owner during 1:1s/ check-in meetings. The risk owner may delegate the monitoring responsibility to someone else, or he/she may have an analytics tool that automatically monitors the situation and flags certain thresholds that need attention. Regardless, the project manager and risk owner should maintain an open line of communication about the risk until it is no longer a potential threat to the project.

The definition of **resolution** in terms of risk management means the either risk trigger has passed, and the risk did not realize; or the risk did realize, and it is now an issue. For risks that become issues, the project manager will carry the risk assessment information to the next step of the GRIDALL (see Chapter 17).

After the risk is resolved, the PM should update the risk register and any reports that contains the risk.

16.3 BEST PRACTICES FOR KEEPING THE RISK MANAGEMENT PROCESS FINELY TUNED

The steps above should enable you to create a RMP that suites the needs of your project and for the company. After the plan is set up and the team understands the process for managing risks, the operation of the plan becomes fairly routine. Here are a few tips for keeping the risk register updated and the risk management process finely tuned:

- Assess your risk register regularly, setting a biweekly or monthly cadence
- Evaluate the risk register when project progress varies significantly from the plan or major changes to the plan are made to ensure there are no "knock-on" risks
- Remember to convert "Risks" to "Issues" as needed
- Continually remind team members to share risks with the project manager and project leader

- Evaluate the success of implemented risk action plans to gather lessons learned
- More broadly, check the effectiveness of the risk management process to see whether the RMP needs to be updated

16.4 SUMMARY

Risk is inherent in the forward-looking nature of goals, and this is especially true in the complex and unpredictable world of biological systems and pharmaceutical development. Some risks in drug development are inherent to the scientific investigation of drug safety and efficacy. These "design" risks should be accounted for in the program valuation by downward adjustment of the NPV. Other risks are due to challenges with implementation of the plan. For these, project managers can use a four-step process of identifying, assessing, responding to, and monitoring risks. It is important for the project manager to keep the risk management process fit for the project and company, sometimes requiring modification to stay relevant and useful.

17 Issues

Joseph P. Stalder
Groundswell Pharma Consulting

CONTENTS

17.1 INTRODUCTION

PMI's PMBOK (7/e) defines an issue as "a current condition or situation that may have an impact on the project objectives". I like to take this definition one step further by saying that an **issue** is an event that has already occurred and has impacted or is currently impacting a goal. In other words, it is something that disrupts your plan, whether it be your schedule, cost, or quality plan for the project.

In the GRIDALL framework, issue detection and assessment are required before a decision can be made on how to resolve the issue. The steps outlined below will help the PM set up a process that aims to course-correct problems that derail the project by focusing the team on enabling the right decision.

17.2 A THREE-STEP PROCESS FOR MANAGING ISSUES

As with risks, the PM is integral to the issue management process. Not only should the PM be concerned with issues that affect the project's overall goals, but also the PM is responsible for applying the methodology for managing issues and the process for resolving issues that will be used throughout the project team structure.

I have found the following three-step process to be effective for managing issues (Figure 17.1):

DOI: 10.1201/9781003226857-20

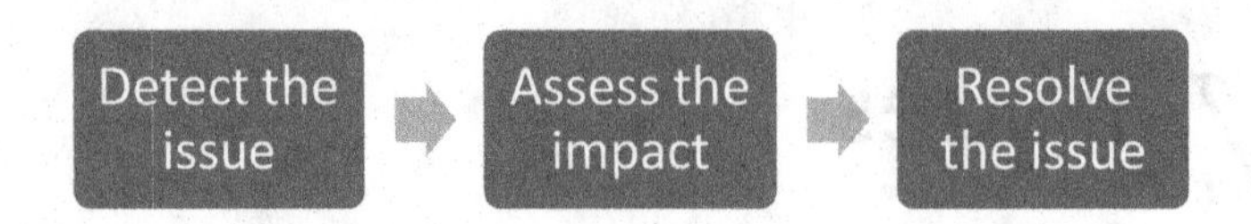

FIGURE 17.1 Three-step process for managing issues.

The sections below will walk through each step of this process. Using the concepts outlined here, you will then be able add issue management to the Risk Management Plan that you created in the previous step of the GRIDALL process.

17.2.1 Detect the Issue

As with risk management, issue detection is the cornerstone of effective issue management. Ideally, an issue would be fairly easy to detect because it was previously identified as a risk and was being monitored by the risk owner. In reality, however, not all issues can be anticipated. Let's look at each scenario.

17.2.1.1 Detecting an Issue That Was Previously Identified as a Risk

Recall that part of the risk management process is to assign a risk owner. This individual is accountable for monitoring the risk and notifying the team if it occurs. The method of detection will depend on the type of risk. For example,

- **Event-driven**: A competitor releases better-than-expected data that diminish the value of the TPP for your development asset.
- **Event-driven**: A health authority provides feedback that the proposed clinical plan is insufficient for approval.
- **Threshold-driven**: Enrollment in a pivotal clinical trial is 75% of planned.
- **Threshold-driven**: A desired specification is not reached in 50% of manufacturing runs.

In many ways, issues converted from risks are easier to manage than unanticipated issues because the risk assessment that was previously conducted can be repurposed for the issue assessment, and the action plan that was put in place can be executed, especially if a contingency plan was already approved by the appropriate governing body. Thus, it becomes much faster to get the team to agree to the recommended action and the decision path is much simpler for risk-converted issues.

17.2.1.2 Detecting an Issue That Was Not Previously Identified as a Risk

As noted in Chapter 16 on risk management, all team members and stakeholders are responsible for detecting risks. This is also true for detecting issues. Also like the risk management process, the project manager should be aware of the same emotional barriers come into play if the person who detected the issue feels uncomfortable raising it to the rest of the team. The project manager and project leader should try to protect the psychological safety of those who bring up issues by both treating them seriously and by recognizing the importance of addressing the issue so the project can carry on toward the goal.

17.2.1.3 Recording the Issue

Once detected, the issue should be recorded on an issue log. Many issue log templates you will see have redundant information with the risk register. However, in the GRIDALL framework, there is no need to repeat information because you can link the issue to a previously identified risk. In addition, the issue log template below contains fields to add a recommendation, discussion points, and the decision path for the issue that you may not see in other issue logs. These fields set up the subsequent step in GRIDALL – the decision.

Background	Related risk	Issue	Recommendation	Discussion points	Decision path
What goal does this issue relate to? What decision was made previously?	*If the issues was previously identified as a risk, what was that risk?*	*What is the current concern/ threat to the goal?*	*What is the team's proposed plan to resolve the issue?*	*What alternatives were considered when evaluating possible paths forward? What assumptions were made when forming the recommended proposal and what is the level of certainty in those assumptions?*	*What endorsements/ decisions will be needed to enact the plan?*

17.2.2 Assess the Impact

Similar to what we saw with risk management, the next step is to assess the impact of the issue on the project's goals. In essence, we need to figure out "how big of a problem is this?" using a consistent methodology. Unlike the risk management step, we do not need to worry about the probability because the event has already occurred (i.e., the probability is 100%), so we just need to assess the impact on the goal.

Again, ideally an issue would have been anticipated by the team as part of risk management. If so, the project manager can simply use the impact assessment that was used during risk management. If not, the team will need to come together to understand the cross-functional implications of the issue. Common impact considerations are cost, time, and strategic value, and some issues may affect one or more of these factors.

17.2.3 Resolve the Issue

Some issues can be resolved by the project team, whereas others will need to be escalated to higher levels of the organization. The approach to resolution will depend on the impact of the issue, and having a documented set of escalation criteria will help guide the project manager to the level of the organization that the issue should be raised to. An example of such criteria is below:

Governing body	Cost	Time	Strategic value
Core team	Up to $250,000	Up to 1 month delay	Meaningful change in
R&D committee	Up to $1M	Up to 3 months	• Competitive landscape
Executive committee	More than $1M	More than 3 months	• TPP
			• PTRS
			• eNPV, ROI, IRR

The table above will vary from company to company and even within projects at the same company. Also, as we will see in Chapter 18, organizations with a two-track governance model will have two sets of escalation criteria: one for high-priority projects and one for standard projects.

17.2.3.1 Decide at Project Team

For issues that are within the project team's control and for which the cost of the proposed response is within the team's decision authority, the decision should be made at the core team level. A useful framework for decision-making is explained in Chapter 18. The project manager is responsible for driving the decision-making process at the core team, and the project leader is responsible for making the decision.

17.2.3.2 Escalate to Higher Level of the Organization

When an issue arises that is outside the project team's control and the cost of the proposed response is beyond the team's decision authority, it will need to be escalated. **Escalation** is the process of shifting ownership of an issue to a higher level of the organization where it is more effectively managed. The project manager needs to know the organization's governance structure to be able to guide the team to the appropriate governing body to resolve the issue. The project manager is also responsible for coordinating the development of the team's recommended response to the issue that will be proposed to the governing body.

17.3 FRAMING THE ISSUE

When bringing an issue to a decision-maker for resolution, it is essential to frame it in such a way that the decision actually addresses the correct issue. I have often seen a decision get made for the wrong issue. In a classic allegory, a carrot producer asked his team how best to get rid of the carrots that were not fit for grocery stores because they were bent, forked, or twisted. Some people proposed selling them to livestock farmers as feed. Some proposed to sell them to soup or juice processors. All of these solutions dramatically undervalued the market price of the carrots. Then, someone proposed instead a new challenge: how to maximize the value of these wonky carrots. The resulting solution: chop them into smaller pieces and sell them in grocery stores as "baby carrots". This solution reduced the carrot farmer's waste and preserved (even increased) the value of the wonky carrots.

The wrong frame can drive the team to the wrong answer. This happens at all levels of business. Even at FDA advisory committee meetings, the committee begins to address the initial question posed to them, only to be later found out that they are going down the wrong path and they need to revise the question (see, for example, the AdComm for COVID-19 booster decision Sep 17, 2021).

Case in Point: Framing the Issue at FDA Advisory Committee

At 167th meeting of the Vaccines and Related Biological Products Advisory Committee, the VRBPAC was asked to vote on the following question:

1. Do the safety and effectiveness data from clinical trial C4591001 support approval of a COMIRNATY booster dose administered at least 6 months after completion of the primary series for use in individuals 16 years of age and older? Please vote Yes or No.

The results of the vote were as follows: Yes = 2, No = 16.

Thus, the committee voted against approval of a booster dose for individuals 16 years of age and older. Committee members expressed concern regarding uncertainties about the benefit afforded by a booster dose relative to the benefit provided by previous vaccination with the primary series in persons 16 years of age and older. Concerns were expressed about post-authorization data demonstrating increased risks of myocarditis and pericarditis, particularly within 7 days following the second dose of COMIRNATY with the highest risk in males 16 to 17 years of age. At the time, there was limited data on whether this risk may be further increased after a booster dose of COMIRNATY. Some committee members also noted the absence of robust data regarding the effectiveness of a booster dose against the then circulating delta variant of SARS-COV-2.

Considering the committee's feedback, the FDA then asked the committee members to vote on a newly formed question that aimed to tease out the concerns expressed and to remove the criteria for "approval". The question then became: Based on the totality of scientific evidence available, including the safety and effectiveness data from clinical trial C4591001, do the known and potential benefits outweigh the known and potential risks of a COMIRNATY booster dose administered at least 6 months after completion of the primary series for use in individuals 65 years of age and older? Please vote Yes or No.

The results of the vote were as follows: Yes = 18, No = 0.

Thus, the committee expressed support for the use of the vaccine in individuals 65 years of age and older because they are at higher risk of complications due to COVID-19 infection.

Thus, with a simple reframing of the question, the FDA was able to gain insight from the advisory committee that was not apparent in the original question.

In the context of the GRIDALL framework, issues can usually be framed in the context of the goal that is being affected. For example, a goal of providing a top-line readout for a clinical trial may be affected by slower-than-expected enrollment into the trial. One way to frame the issue is "how can we increase enrollment?". However, this may miss some potential solutions. If the issue is framed as "how can we achieve top-line readout at the same time without jeopardizing the quality of the data?", you may get solutions such as "assess the likelihood of success of evaluating the study with fewer total subjects enrolled" or "evaluate the addition of real-world data to the existing dataset to support analysis".

17.4 SUMMARY

In the context of the GRIDALL framework, an issue is an event that disrupts the plan and prevents the team from achieving the project goal. An issue can be the result of a risk being realized or from an unexpected event that jeopardizes the project goal. Issues that were previously identified as risks are usually easier to manage because the risk management process had already assessed and planned a response to the event. Regardless of the origin of the issue, it will have to be resolved, either through a decision made at the project team or by escalating to a higher-level part of the organization. When working up a proposed solution to the issue, it is important to correctly frame the issue so that the best possible solution is selected.

18 Decisions

Joseph P. Stalder
Groundswell Pharma Consulting

CONTENTS

18.1 INTRODUCTION

A **decision** is a commitment made by an organization to use resources – money, energy, or both – for a directed action. In the GRIDALL framework, a decision is a commitment to use the organization's resources to address an issue. There are hundreds if not thousands of decisions made along the drug development journey, not least of which is the initial decision to invest in the development opportunity; however, the types of decisions that pertain to GRIDALL are those that resolve threats and exploit opportunities to the Clinical Development Plan. That will be the focus of this chapter, whereas investment decisions are the focus of Chapter 5 on portfolio management.

Note the distinction between *decisions* and *endorsements*. Decisions commit resources, whereas endorsements provide agreement, such as to a strategy or to a change in goals. A **governance structure** at a large company may have several governing bodies that review certain aspects of a proposal before a decision is made to commit resources. These review bodies provide endorsements, whereas there should only be one governing body that can make a decision. I like to describe review bodies as "forums" and decision-making bodies as "committees". Thus, *forums* provide endorsements, *committees* provide decisions.

While every team member desires fast and well-made decisions, project managers particularly are keen to obtain decisions so that the project can move forward. Project managers support the project and the project team by helping to navigate through the organization's governance structure to get decisions made. Therefore, project managers need to thoroughly understand general decision-making principles as well as the organization's specific decision-making process.

DOI: 10.1201/9781003226857-21

This chapter will describe the theory around decision-making, including principles of well-made decisions. We will then describe some governance structures that enable fast and efficient decision-making. Finally, we will walk through a five-step decision-making process that a PM can use to drive decisions for the project team.

18.2 PRINCIPLES OF DECISION-MAKING

Decision theory is complicated, and countless business management books have tried to tackle the topic. A very simplified theory of decision-making is that the best decision is made when the right person with the right authority is given the right information (Figure 18.1).

In practice, however, decisions are confounded by constraints, personalities, biases, opinions, politics, and so on. In addition, the idealized formula above is complicated by the fact that determining the right person, the right authority, and the right information is subjective and often controversial.

What is meant by the "right person"? Rarely is there a single person who has all the experience and knowledge to make decisions across the entire drug development spectrum. Hence, the formation of groups of people that collectively provide a wide range of experience and knowledge. These subject matter experts are there to provide input and perspective to the decision-maker (usually the committee chairperson) so that the right decision can be made. The chairperson is usually authorized to make decisions within certain budgetary and resource assignment parameters, as set out by the committee charter.

What is meant by the "right authority"? The most effective companies have some sort of policy that guides decision-making, because *policy takes the politics out of the process.* I say "some sort of policy" because policies are usually meant to control specific situations; however, with decision-making, it is not possible to ascribe a policy to the myriad issues that may come up. Thus, most companies will create a **governance model** that describes the structure, remits, and roles for each governing body in an organization. Project managers are then able to use this model to navigate the subjective complexities and obtain decisions that stick.

What is meant by the "right information"? In drug development, we are often required to make decisions based on incomplete information. Obtaining the right information becomes a balancing act between being able to move the project forward quickly and waiting for more data to make a more informed decision. There is no clear answer, and that is why our decision-makers are so important for being able to

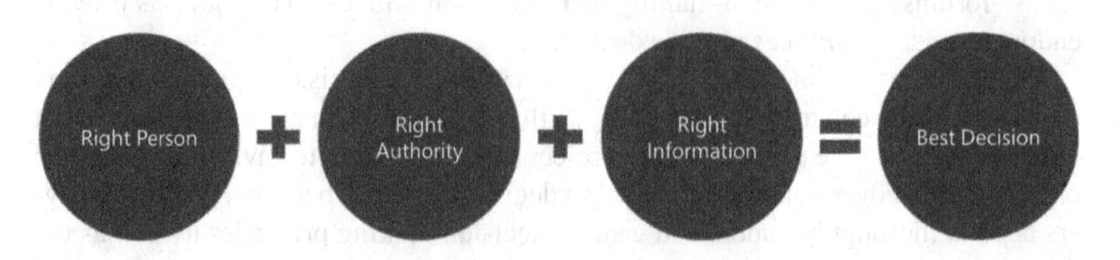

FIGURE 18.1 Theoretical representation of the ideal decision-making situation.

use their experience, intuition, and judgment to determine how much information is needed for them to comfortably make the right decision.

What is meant by the "best decision"? Sometimes, the outcome of a decision is not what we expected or hoped for, and some might deem it in the end to have been wrong decision. However, a wrong decision can still be the best decision at the time it was made because it seems right based on the information available at the time.

Instead of focusing on right and wrong decisions, a PM should focus making well-made decisions. The "well-made" decision has the following qualities:

- The decision is made within the overall context of the company's strategy, mission, values, and ethical standards.
- The decision is *sound* based on available information, *considerate* of the perspectives of relevant stakeholders, *prudent* of both the present and long term, *circumspect* to the overall project instead of any individual or functional view, and *balanced* with regard to risks and rewards.
- The decision is made by a single, accountable individual at the lowest appropriate organizational level.
- The decision is made using a defined decision-making process, in a timely manner, and communicated to the appropriate stakeholders.
- The decision is supported and only revisited when new information becomes available.

An ill-made decision is one that does not follow decision-making principles, irrespective of the desired outcome. To create a culture of psychological safety where decision-makers feel comfortable making decisions, leaders should acknowledge and evaluate "wrong" decisions but not apply disciplinary consequences. However, "ill-made" decisions should be addressed and remediated to avoid future mistakes.

18.3 GOVERNANCE STRUCTURES

Whether defined implicitly or explicitly, every organization has a governance structure. In a small company with few organizational layers, most decisions are likely made by senior management, oftentimes with the CEO serving as the sole decision-maker, forming a top-down decision process. In larger companies with many organizational layers, decision-making rights are typically dispersed throughout the organization, forming a matrix-style decision process.

There are advantages and disadvantages to each approach. Companies with top-down management are often fast at making decisions, but they may not have the benefit of comprehensive experience and information to make the right decision, and this model quickly becomes ineffective as a company grows in size and complexity due to the sheer volume of decisions that are needed. On the other hand, companies with matrix management can benefit from broader experience and information, but they can be hindered by the complex process of obtaining a decision that needs several reviews and endorsements before the ultimate decision is made.

The best approach may depend on the type of project. If it is a promising high-priority project that needs to move fast, a fast-track pathway similar to the small

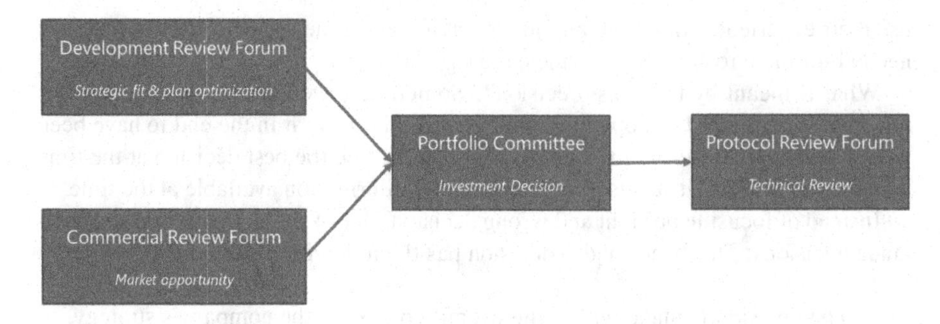

FIGURE 18.2 Example governance structure for a mid-sized company that enables fast, effective decision making.

company top-down approach will be best. In this model, the project team's spend authority is increased, and the team gets prioritized access to higher levels of the organization for issue escalation. If it is a standard priority project, it will go through the standard matrix decision pathway. Organizations that set up this two-track system (i.e., fast track for promising projects, standard track for all the others) will be able to take advantage of nimble decision-making for high-priority projects and the scalable, predictable approach for standard projects.

Company culture also plays a part of how decisions are made. I have experienced small companies where the leaders decry "governance" as being too bureaucratic; they argue that governance kills the nimble spirit of biotech (ironically, these leaders that claim to be "biotech" do not realize that their company is not actually using biotechnology and that the reference to "biotech" as being a small, scrappy, nimble company is no longer relevant now that some true biotech companies have become large, multinational enterprises). I have also seen large companies that have poorly defined governance structures where decision-making is confusing and chaotic. On the flip side, I have seen companies that embrace the concept of being a decision-centric organization, where the governance is simple, the decision roles are clear, and decision-makers are readily accessible. Such a structure for a mid-sized or large company might look like this (Figure 18.2).

Because project managers are instrumental in driving decisions for the project team, we need to know both the formal decision structure and the informal decision culture at our companies. At small companies, it is important to realize the experiential limitations of the top-down management approach and to vouch for external consultation when appropriate. At larger companies, it is important to understand the matrix management environment to be able to quickly navigate the governance structure and enable for fast decisions.

18.4 A FIVE-STEP PROCESS FOR EFFICIENT DECISION-MAKING

After the team realizes they have an issue to deal with, it is easy for the team to feel lost without a clear project plan. This is the project manager's opportunity to shine. By setting up a path to resolve the issue, the PM will guide the team through the uncertainty and bring the project back on track. I have found the following stepwise

FIGURE 18.3 Five-step process for efficient decision making.

approach to be most efficient in obtaining a decision that resolves the issue. Sharing this process with the team gives them comfort in knowing there is a path toward getting the issue resolved and getting the project moving again (Figure 18.3).

We will go through each step in detail below, highlighting the use case for the types of decisions that pertain to the GRIDALL framework.

18.4.1 Set the Context

The first step is to frame the issue within the context of the broader situation. For a new opportunity investment decision, the context could be that there is an unmedical need and there are data to suggest that your drug might be able to address that need. For a change to development plan (i.e., because there is an issue with the plan), the context could be that a certain set of assumptions were made at the outset of the project, but new information has invalidated an assumption and a new plan needs to be put in place. For a stage-gate transition decision, the context is that data were obtained from the previous stage to support either continuing or terminating further development.

Decision-makers or executive sponsors should communicate the strategic context when requesting information to support a decision, and the PM and PL should seek clarification if the strategic context has not been provided completely or clearly. When doing so, it is helpful to consider the following (Figure 18.4):

Define the Issue	Assess the Urgency	Understand the Constraints
• Frame the issue as a challenge to the current plan/ goals • Was the issue a previously known risk? • What decisions were previously made that led to the issue? • Is the problem the result of changed assumptions or information that now requires a change of course?	• When does the decision have to be made so as to not affect downstream events? • Are all the right people on the Core Team to make the decision, or does the need to be escalated?	• What are the key project constraints with respect to time, cost, quality? • What interdependencies exist with other parts of the project? • Are there criteria by which success or failure will be evaluated?

FIGURE 18.4 Considerations for setting the context of a decision.

18.4.2 Define the Roles

Once the issue is framed up within the appropriate strategic context, the next step is for the PM to define the roles involved in the decision. In any size company, even clear, well-framed decisions can be derailed by uncertainty over who is the authorized decision-maker. This is especially apparent in larger companies with matrix management and many governing bodies because it can be difficult to know where to go for a particular decision. In these environments, the PM would benefit from deploying a **decision-making framework** that will help to reduce confusion. There are many decision-making frameworks available to clarify decision accountabilities. Here are a few:

- DAI (decision-maker, advice giver, informed stakeholder)
- DACI (driver, approver, contributor, informed)
- DARE (decider, advisor, recommender, execution stakeholders
- RAPID (a loose acronym for recommend, input, agree, decide, and perform)

The best decision-making frameworks empower a single accountable person to make the decision. For operational decisions, this is the person who has the appropriate information, experience, and expertise to make the decision that best resolves the issue. For strategic decisions related to investment opportunities, stage-gate transitions, and changes to the clinical development plan, this person is often the chairperson of a governance committee.

All other relevant players in the decision-making process are considered stakeholders. A **stakeholder** is a person, function, or team that contributes to, is affected by, or is responsible for executing a decision. Stakeholders can come from many parts of the organization. To identify relevant stakeholders, the PM should consider the following:

- Those who are affected by the decision and its subsequent actions, by either direct responsibility or indirectly as a functional area head that provides resources,
- Those with information and expertise in the area who can provide input to enable the best decision, and
- Those whose understanding and support of the decision is important for successful implementation.

By defining the roles before making the decision, the project team can consult the appropriate people and review forums prior to the decision being made. This reduces the swirl we often see when, after a decision is made, a person will object to the decision based on some ground that may not have been considered. If the decision path is clearly laid out and everyone knows how the decision is going to be made, the chance of swirl is minimized.

18.4.3 Gather Information

The next step is for the PM to work with the team to amass the information needed to enable a decision. This information should be organized into a storyline and proposal that the decision-maker can use to make the decision. This step of the process can easily get out of control and put the team in analysis paralysis. To stay focused, the PM will guide the team to only gather the information that is necessary to support the decision-maker's choice. The PM should use the strategic context defined in the first step to keep the information gathering at an appropriate level.

For new investment decisions, it is helpful to think of the lack of commitment to the opportunity as an issue that is preventing the team from maximizing the value of the asset. The storyline would go something like this:

1. Our current development plan is missing out on potential value.
2. Here is a new opportunity that we think will provide value.
3. We now request resources to execute the first stage of the plan.

For stage-gate transition decisions, the storyline would go something like this:

1. We promised we would deliver data to support a future investment decision.
2. Here is the data and the recommendation to move to the next stage.
3. We now request resources to execute the next stage of the plan.

I find it helpful to summarize the key assumptions of the investment proposal into a Governance Summary slide that contains the following information:

Project attribute		**Description/value**
Study design	Study description	Phase, randomized, double-blind, placebo-controlled, treatment, population etc. n=xxx
	Endpoints	1: xxx; Key 2: xxx
Budget	Cost to next stage gate	$Xm (internal/external)
	Total project cost	$Xm (internal/external)
Timeline	First subject dosed	MMM YYYY
	Data readout	MMM YYYY
	Launch	MMM YYYY
Risks	PTS	X%
	PRS	Y%
Valuation	PYS	$Xm (YYYY)
	eNPV	$Xm

For plan-change decisions, I have found it helpful to prepare a slide that summarizes the key information as follows:

Background	What goal does this issue relate to? What decision was made previously (e.g., by other governing bodies)?
Issue	What is the current concern/ threat to the goal?
Discussion points	What alternatives were considered when evaluating possible paths forward? What assumptions were made when forming the recommended proposal and what is the level of certainty in those assumptions?
Recommendation	What is the team's proposed plan to resolve the issue?
Next steps	What additional endorsements/decisions will be needed to execute the plan?

When collecting this information, it is helpful to consider the following questions:

- What options and scenarios are available to address the issue?
- Is data or literature available to support the decision (e.g., analogous clinical data, enrollment metrics from prior study)?
- What SOPs/ policies, regulatory/legal requirements, guidelines/best practices/lessons learned may apply?
- Is there anyone within the organization who has knowledge/experience with the situation that you can consult for advice (e.g., senior leaders, subject matter experts)?
- Is there anyone outside the organization you can consult for advice (e.g., vendors, consultants, KOLs, steering committee)?

After the right information is gathered, it should be organized in a presentable format that tells a story. I often spend time with the team to create a storyboard before putting slides together. The storyboard then helps frame the content that needs to be included, keeping in mind that some content is intended for the pre-read and not for the presentation. The pre-read can be shared in advance of the meeting so that attendees can read the background information and the team does not need to spend time reviewing it during the meeting. The pre-read slides are then removed from the presentation (aka showfile) deck so that only slides that are relevant to the issue are presented. General rule of thumb is six to eight slides for a 1-hour meeting.

18.4.4 Make the Decision

In this step, the project manager ensures the right information is provided to the decision-maker to enable a decision. This usually involves working with the committee's secretariat to schedule the presentation, ensuring the pre-read is available in time, and preparing the final showfile that will be used in the meeting. During the meeting, the project manager may be asked to present parts of the presentation, but it is usually the project leader who is responsible for conveying the proposal. The project manager and project leader may want to hold a pre-meeting with the committee chairperson to introduce him/her to the issue and answer any initial questions about the issue.

During the committee meeting, the chairperson will drive the flow of the discussion. This includes

- Clearly stating the questions that are needed to be answered.
- Considering all options that were identified and the implications for each option, including short- and long-term costs and benefits.
- Validating or challenging assumptions.
- Choosing the optimal option based on the available information and relative to the stated objectives for the project.
- Defining circumstances under which the decision should be revisited.

Before closing the meeting, the chairperson should recap the questions that were needed to be answered and the committee's decision for each question. The team should also ensure that the decisions and next steps are understood and ask any clarifying questions if the path forward is not clear.

18.4.5 Document and Implement

Decisions made without adequate documentation and communication to the implementers (referred to as *Performers* in the RAPID model) become subject to being changed, overturned, or simply not acted upon. Therefore, the project manager should memorialize the decision in clear and definitive terms and in a way that inspires the implementers to carry out the subsequent actions. It is helpful to keep a running log of decisions made so that future team members can be properly informed of the decisions that were made. The table below gives an example of a decision log. It may look very similar to others you have seen but note the reference to the issue that connects the decision to previous component of the GRIDALL framework. By including a description of the issue, the context of the decision will become clearer.

Issue	Decision	Rationale	Next steps	Decision-maker	Decision date
What was the problem that needed to be resolved?	What commitment was made?	Why was this choice made among all the other alternatives?	What are the downstream activities needed to execute the decision?	Who made the decision?	When was the decision made?

Implementation of the decision requires the functional heads to allocate resources (people and money) and the project manager to convert the decision into an action plan that describes who will do what by when. The transition to implementers can be made more effective by involving the "doers" early and jointly developing the implementation plan. The implementation plan should include a clear transfer of accountability, and the project manager should establish the appropriate channels to monitor progress and help remove barriers when needed.

18.5 SUMMARY

Decisions are needed to resolve issues that arise during the execution of a project plan. The right person with the right information and the right authority is in the best position to make the decision, and project managers can facilitate this concept by outlining the decision-making process as soon as an issue is identified. Project managers are especially adept at supporting the project team to navigate the organization's governance structure to make sure decisions are made quickly and they last, meaning there is no swirl after a decision is made. Project managers can use a five-step process of setting the context, defining the roles, gathering information, making the decision, and documenting and implementing the decision.

19 Actions

Joseph P. Stalder
Groundswell Pharma Consulting

CONTENTS

19.1 INTRODUCTION

Every decision is followed by an action. In the GRIDALL framework, an action is an activity that completes the work to implement a decision. Implementation requires resources either in the form of people's time or the organization's money, and it is one of the project manager's key responsibilities to ensure resources are being used with maximum efficiency. Therefore, the project manager needs to know how to identify, define, record, and monitor action items. These steps are described below.

19.2 PRINCIPLES OF MANAGING ACTIONS

Project managers should know some of the key principles of managing action items and the implications of these principles on the role of the Project Manager.

Principle	Implication for Project Managers
An **activity** is something that needs to be done; a **task** is an activity that a person needs to do	Create tasks, not activities
What gets assigned gets done	Don't be afraid to assign a person, if they don't think they're the right person – let them say so
Diffusion of responsibility → confusion of ownership → no one takes action	Assign actions to an individual, not to a group
Accountable ≠ responsible	Often people at the Core Team level are **accountable**, not necessarily **responsible**

DOI: 10.1201/9781003226857-22

In the common workflow, a decision made at a governance meeting will not explicitly state the action item to be completed. Although the RAPID framework will identify the individual or group that will **perform** the recommended action, it is typically up to the Core Team to create the action plan needed to implement the decision. Therefore, a project manager will bring the decision to a Core Team meeting and create an action plan on how to make it happen. First, a project manager may want to list the activities that need to be completed. During the meeting, he/she can assign those activities to the appropriate core team representative in the room. This does not always mean that the core team representative will be responsible for doing the task, but that he/she needs to see to it that it is done either through delegation or outsourcing. The project manager should be clear on who owns the action item so that there is clear accountability.

19.3 A FOUR-STEP PROCESS FOR MANAGING ACTION ITEMS

Managing actions is a project manager's forte. We explored in Chapter 9 the digital tools the project managers can use to plan, track, and manage actions. Yet it may take some work to create a clear action plan with the team to ensure the decision is clearly understood and the action will deliver on the commitment that was made. A project manager can use the following four-step process to manage action items.

19.3.1 Identify

Project managers need to have the situational awareness to be able to recognize when an action is needed. In the GRIDALL framework, actions are identified in the decision log as "next steps", but there are other sources of actions that may be harder to tease out. For example, in a meeting where a topic for decision seems to be swirling, there is likely an action to gather more information to inform the decision. In a meeting to inform the rest of the team of something, there may be an action for a sub-team or task force to make recommendation on how to resolve an issue. The project manager should always be listening for these cues and help the team figure out what action needs to be taken.

19.3.2 Define

Figuring out what action needs to be taken is arguably the most difficult yet important step in the process. The project manager should help the team clearly define the "who, what, and when" of an action plan.

- **Who** – Each action should be assigned to a single individual, and that person should confirm his/her acceptance and understanding of the action item before the meeting is adjourned. Determining the most appropriate person is usually obvious (in which case I like to identify him/her as the *Performer* in the RAPID model), but sometimes a task may need multi-functional attention, so the project manager will need to ask the team if someone is willing to own the action. Recall that the owner is not always responsible for performing the task, but he/she is accountable for ensuring the action is completed.

- **What** – Each action should clearly define the expected deliverable(s). For example, the deliverable could be a document, a presentation, a recommendation, or a piece of information.
- **When** – Each action should have a due date. Assign a due date that takes into consideration milestones, handoffs, and triggers for downstream activities.

I find it helpful to use the last 2–3 minutes of a meeting to recap the actions and to get verbal confirmation from the action owners that they agree to and understand the action. There is a special caveat to this verbal confirmation for international teams where a "yes" may have a different cultural meaning than what you might expect. Chapter 7 (Managing International Projects) has more details on this.

If an action plan cannot be established during the meeting either because the right people are not in the room or because the plan is complex and requires more input, set up a follow-up call with the appropriate stakeholders to work up the plan. I call these following-up sessions "plan the plan" meetings.

19.3.3 Record

Recording the action not only establishes a historical reference, but also helps with dissemination of the action to the appropriate people. We mentioned previously that the person assigned to own the action may not necessarily be the one to carry out the work; he or she may delegate the work to others. In this case, the person(s) responsible for doing the work will need to understand the context and rationale for the action. The action log template below describes the action as it relates to the issue that needed to be addressed and the decision that was made. The action item owner, then, can use this context when describing the action to his/her delegates.

Context	Action	Owner	Due date	Check-in timing
What was the problem that needed to be resolved? What decision was made, and what was the rationale for that decision?	What downstream activities are needed to execute the decision? What deliverable/output is expected?	Who will take action and report on status?	When is the deliverable/ output due?	When will the owner report back on progress?

I find the most useful action item trackers are those that everyone on the team can access and update. There are many electronic tools for creating a shared system for tracking tasks, like spreadsheets, to-do lists, and task management solutions. Chapter 9 has more detail on the utility of these tools. Note that the fields in the above action item tracker are slightly different than what most task management solutions offer. A task management solution might have fields for a task title, assigned owner, due date, priority, progress, category, etc. They usually do not have a field to include the context of the action, so this will need to be communicated via other means (e.g., verbally or email).

Note that the above template is useful for assigning a task to a single owner, but the actual action plan may involve several people who are involved with completing discrete parts of the action. For these complex actions, you may need a RACI matrix to clarify the roles and responsibilities of the task. It is up to the action owner to determine how best to perform the action, and the project manager can help when needed to clarify and document the plan.

19.3.4 Monitor

The last column in the action log template allows the project manager and action owner to agree on a timeframe to check in on the progress of the action. Check-ins serve as a mechanism for the core team to monitor progress and help remove barriers with needed. While the project manager should be discussing progress and risks with the action owner through regular communication channels (e.g., 1-on-1 meetings, hallway conversations, online chat), the check-in referenced in the template is meant to be the time when the action owner will come back to the core team to discuss progress and risks. The trigger for the check-in may be a data readout, or a milestone event, or delivery of an output from an external stakeholder (e.g., health authority feedback or a report from a CRO).

19.4 IMPOSSIBLE MISSIONS

It happens from time to time that a proposal gets approved, but the implementation plan was not fully considered, and the execution of the decision is found to be impractical. This may be especially true when the proposer has different incentives from the implementers, such as a sales team pitching a service that the operations team then has to deliver. Smart decision-makers think carefully about risks of implementation, with big decisions as well as small ones. They will assess not only the feasibility of the proposal itself, but also the credibility of the people recommending the investment. All stakeholders in the decision process should also consider risks to implementation as well. But what should a delivery team do if they are handed a decision that is impossible to implement?

Part of an action plan may be to return to a governing body to clarify or adjust some project assumptions. Perhaps the team has finally had a chance to work out the action plan, and they identify that it may take more money, time, or resources to deliver. In this case, the team should return to step 1 of the GRIDALL process by setting the new project goal, then documenting the risk, converting it to an issue, and proceeding to the governing body for a new decision. These situations are especially important after a new decision is made because there is a valuable lesson learned that can benefit future projects and the organization at large.

In the competitive environment we work in, leaders are often pushed to make decisions based on less-than-complete information. Assumptions made during the proposal may turn out to be untrue. Perhaps the team has started to implement the decision and found that the delivery target is too ambitious. For example, the team has started a new trial and enrolled 10% of the patients, but they should have enrolled 25% at this point. In this case, the revision to the new action plan will need to be taken back to the governing body to propose a new set of assumptions and a new project goal.

Each company has its own risk tolerance. Sometimes, taking action requires pushing the envelope, taking bold chances, or venturing into uncharted territory. This may make some senior leaders uncomfortable, so the decision may be to start with a small part of the proposed project and check back at a pre-specified decision point to request further funds or resources.

19.5 SUMMARY

Actions produce outputs. In the context of the GRIDALL framework, actions convert decisions into results. Decisions made at governing bodies are typically converted to action plans at the core team, and the project manager is the facilitator of this process. There are some key principles that a project manager should know when managing actions, such as assigning an action to a single accountable person.

Project managers can use a four-step process for managing actions. First, they need to be able to identify when an action is needed. Then, they should define the action by clearly assigning the "who", defining the "what", and agreeing on the "when". The project manager should then record the action in a tracker and possibly use a task management solution to disseminate the action to the team for transparency and monitoring. Finally, the project manager should continue to monitor the progress of the action, using check-ins to gauge whether interventions are needed.

Despite a leader's attempt to make wise choices, there are times when the delivery of an output is simply not possible given the resources committed to it. In these situations, the team should work up a new plan and return to the governing body with better assumptions and plans, cycling back in the GRIDALL process as needed.

20 Lessons Learned

Joseph P. Stalder
Groundswell Pharma Consulting

CONTENTS

20.1 INTRODUCTION

The final component of the GRIDALL framework is arguably the most important for an organization. PMI PMBOK (5/e) defines a lesson learned as "the knowledge gained during a project which shows how project events were addressed or should be addressed in the future with the purpose of improving future performance". Every decision made and subsequent action taken can become a lesson learned, and there is value in capturing and sharing these lessons learned so that other project teams can adopt successful approaches and avoid unsuccessful ones. The successful approaches become best practices, and companies that consistently apply best practices will have a competitive advantage over those that repeat the same mistakes. Furthermore, the compilation of lessons learned and best practices becomes permanent institutional knowledge that is resilient to employee turnover and changes to the organizational structure.

Despite the value of the cultivating an "always learning" mindset, companies often have difficulty capturing lessons learned and making them easily accessible to beneficiaries, and biopharma companies are no exception. Some of the challenges the project managers face when attempting to capture lessons learned include the following:

DOI: 10.1201/9781003226857-23

- Companies with programs that reward only successes unintentionally create a culture where teams are disincentivized to share their failures.
- Team members do not want to spend time reflecting on the past because they are assigned to new projects or tasks immediately that require their attention.
- Senior leaders are not fully committed to supporting the lessons learned processes because they have not seen the value of true knowledge management.

To overcome these challenges, project managers should try to create a team culture that embraces failure as a learning tool and rewards team members for a successful attempt, not just a successful outcome. They should adopt a continuous improvement mindset that aims to capture lessons learned in real time, rather than waiting until the end of the project. And they should demonstrate the value of the lessons learned process by highlighting what was learned from other projects as the team plans for a new project.

20.2 TYPES OF LESSONS LEARNED

There are two ways to capture lessons learned, distinguished by the timing of conducting the lessons learned activity. First is a post-mortem analysis, which occurs after the project has completed. The second is kaizen, or continuous improvement, which occurs mid-project. I find that post-mortems are best for projects that have a long cycle time and occur infrequently, for example, a clinical trial. Continuous improvement processes are best used for projects that have short cycle times and occur frequently, for example, activating an investigative site in a new region.

20.3 A FIVE-STEP PROCESS FOR GATHERING LESSONS LEARNED

I have found the five-step process shown in Figure 20.1 to work well for gathering lessons learned.

20.3.1 Collect Initial Feedback from Stakeholders

The first step is to solicit feedback from project stakeholders on what worked well and what could be improved. It is helpful to use an anonymous survey to collect this initial feedback. At the beginning of that survey, share the context and reason for conducting the lessons learned exercise. You can break the survey into topic areas and include prompts to get the stakeholders thinking about their responses.

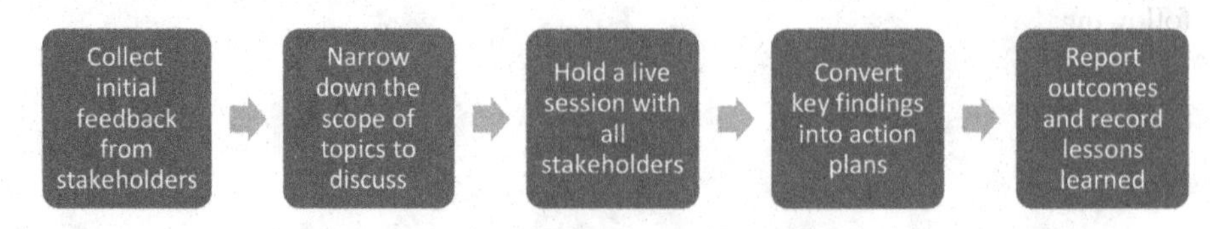

FIGURE 20.1 Five-step process for gathering lessons learned.

Topic Area	Prompting Questions
Project planning	• Was the project plan realistic? • Did the planning effort include key stakeholders? • Were all major activities accounted for? • Did the individual team members responsible for the work provide the estimates? • Were appropriate staff involved in the planning effort?
Project Execution	• Was the project executed according to plan? • Was there an appropriate level of commitment and urgency? • Was change management performed appropriately? How well were scope changes managed? • Was risk management performed appropriately? • Was issue management performed appropriately? • Was the plan flexible enough to adapt to unforeseen circumstances? • Did the project unduly stray from identified goals and objectives?
Staffing, roles and responsibilities	• Were people appropriately assigned to the project? • Was staffing sufficient? • Were roles and responsibilities clear? • Did team members and stakeholders fulfill their roles appropriately? • Were leadership and governance effective? • Were team members appropriately experienced/skilled?
Teamwork	• Did the teams/ sub-teams work effectively together? • Did the teams share a common goal/ mission? • Were all stakeholders appropriately engaged, involved, and available?
Communications	• Were communications effective? • Were meetings effective? • Were expectations clear? • Did you know where to go to get project information? • Did communications go to the right people at the right time? • Were communications with the vendor/ CRO effective? • Did we escalate effectively and at the right time? • Were issues dealt with effectively?

The project manager should send the survey about 3 weeks before the planned meeting, giving stakeholders 2 weeks to respond and giving him/herself 1 week to synthesize responses.

20.3.2 Narrow Down the Scope of Topics to Discuss

After the initial stakeholder feedback is available, the project manager and project leader should read through it and select topics they want to discuss at the live session. In general, topics should be relevant, significant, and mutable.

- **Relevant**: Discuss only things that pertain to the project or activity at hand.
- **Significant**: Discuss things that have impact on the outcome project or activity.

- **Mutable**: Discuss only things that are changeable. For example, you cannot change the fact that a competitor advanced their program quickly and took up all the investigative sites you wanted for your study, but you can change the way your company monitors the competitive landscape to become aware of such events in the future.

This exercise of synthesizing responses usually takes about a week. The output is typically a slide deck that has one slide per topic. The project manager can then send this slide deck to the participants of the live session so that they can prepare for the discussion.

20.3.3 Hold a Live Session with All Stakeholders

Live feedback sessions are still the gold standard for capturing lessons learned, but these meetings need to be facilitated carefully to avoid animosity and finger-pointing. Before starting the discussion, it is important for the project leader to outline the purpose and goals of the discussion and to provide an objective account of the project or activity history, including the goals, risks, issues, decisions, and actions for the project or activity, as well as the external events and strategic context that shaped the project's path.

Also before the discussion, it is important for the facilitator to describe how the meeting will be run. I have found the "taking stack" technique to be very effective in these types of meetings because it ensures everyone who wants to speak to a topic can say what is on his/her mind without being interrupted or challenged by others. Also, the facilitator should lay the ground rules for the discussion, for example,

- be nice,
- do not blame (no names allowed),
- be constructive, and
- focus on solutions rather than rehash the past.

During the discussion, the facilitator should record ideas and findings on a whiteboard to make sure the group agrees to what is being recorded. The facilitator can use this opportunity to identify themes and connect common issues across the topics that were previously selected.

After the open discussion of topics, the facilitator should save time at the end to identify the highest priority findings that need to be addressed. These items will be converted to action plans in the next step.

20.3.4 Convert Key Findings into Action Plans

After the live session, the project manager and project leader should align on the key findings and convert them into recommendations for change. For example, if a root cause of an issue was identified, determine how to avoid that trigger. If a best practice was identified, determine how to exploit it in the future. These recommendations for change should take the form of action plans that define the who, what, and when of implementing the change. Some examples of actions that could result from a lessons learned session include the following:

- establishing corrective and preventative action,
- creating or revising standard operating procedures (SOPs) and other procedural documents,
- identifying ways to integrate the lesson into future practices,
- clarifying roles and responsibilities for certain tasks, and
- establishing new roles to monitor activities that need oversight.

20.3.5 Report Outcomes and Record Lessons Learned

After the live session, the project manager should summarize the lessons learned and resultant actions and communicate the findings to the project team and relevant stakeholders, including functional area heads. For actions that were identified, it then becomes the responsibility of the functional area head to follow up to ensure the action was completed.

Ideally, historical project information and lessons learned would be stored in a way that is easy for other project teams to find the information pertinent to their specific situation. I find the situation-solution-outcome-recommendation (SSOR) framework to work best for capturing lessons learned.

Situation	Solution	Outcome	Recommendation
What was the goal? Was the situation recognized as a risk? How did the risk become an issue?	What decisions and actions were enacted to address the issue?	How well did the solution resolve the issue? To what extent did the solution meet its goal? If it fell short of the goal, why?	Would you enact the same solution again next time?

The recorded lessons learned then become project memory and institutional knowledge. Institutional knowledge is not subject to employee turnover or changes in the organizational structure. The next step is to get that institutional knowledge to the hands of the people who need it.

20.4 HOW TO IMPLEMENT LESSONS LEARNED AND BEST PRACTICES

Lessons learned and best practices are only valuable if they are surfaced at the right time and adopted by future teams that are approaching a similar situation. This is how the organization gains benefit from the lessons learned process, and the feedback loop is a key concept of the GRIDALL framework. Lessons learned can be applied to future goals, risks, issues, decisions, and actions. There are several ways to implement lessons learned and best practices.

- **Build a lessons learned library**: The simplest and most common approach is to house lessons learned on a central list or spreadsheet so that other team members or project managers can search through. The downside of this approach is that it requires team members to go to the library and look for the appropriate content, rather than having the information pushed to them at the right time. Another downside is that there may be several lessons related to the same situation, and the learner must read through them all to form a best practice.

- **Build a knowledge base**: A more comprehensive approach to recording lessons learned is to create a knowledge base of best practices. A **knowledge base** is an online repository of curated information that can be drawn upon by a user. A best-practice knowledge base can be in the form of a wiki-style collection of articles that describes how to approach processes in a narrative form and link to related articles for the learner to gain more insights. For example, a lesson learned on creating a target product profile may link to a page that describes how to create a probability of technical success.
- **Establish a network of subject matter experts**: Another way to take advantage of lessons learned and best practices is to establish subject matter experts who have deep experience and expertise in a particular situation. This person becomes the "go to" for guiding others through that situation.
- **Automated just-in-time training**: The most sophisticated yet effective approach is to identify tasks in the project schedule that relate to trainings available in your organization's learning management system (LMS). Then, when the task is approaching, the LMS will trigger a training requirement for those assigned to the task. For example, the primary completion date for a clinical trial is approaching, and the team needs to be trained on the SOPs and work instructions that govern the database lock and unblinding activities. The task in the project schedule will transfer the task name, dates, and resource names over to the LMS, which then assigns the relevant trainings to the involved team members with the appropriate due date.

20.5 SUMMARY

Every action provides an opportunity to learn a lesson, and lessons that become best practices can become a competitive advantage for an organization. Approaches to gathering lessons learned include post-mortems, done after a project is completed, and continuous improvement, done while the project is still in flight.

Project managers can use a five-step process for gathering lessons learned. First, a project manager can collect feedback from stakeholders. Then, the project manager and sponsor can narrow down the scope of topics to discuss at a live discussion session. The project manager and sponsor will then hold a meeting to discuss the feedback and elicit further information to help them identify what worked well and what needed improvement. After the live session, the project manager and sponsor will convert the key findings into action plans. Finally, the project manager should record the lessons learned, using the situation–solution–outcome–recommendation model as the preferred format.

A key concept of the GRIDALL framework is to apply the lessons learned to future goals, risks, issues, decisions, and actions. There are several ways to make lessons learned and best practices available to the rest of the organization, including a lessons learned library, a knowledge base, a network of subject matter experts, and automated just-in-time training.

Appendix I

Team Performance Survey

Mission, Vision, Roles, & Responsibilities

1. Our team goals and objectives are clear.
2. The team is delivering results according to the plan.
3. I feel energized by the challenge the team has agreed to take on.
4. I understand the roles and responsibilities of the team members.
5. The team understands my role and responsibilities.
6. Team members act consistently with their agreed and understood roles.

Operations

7. Our team has agreed team norms that are honored and regularly reviewed.
8. We are communicating openly and in a timely manner.
9. Significant issues or decisions are dealt with quickly and directly.
10. Tools and processes are in place to effectively support teamwork.
11. Meetings are a good use of my time.
12. Our team works well both during and between meetings.

Decision Making

13. Our team makes decisions effectively.
14. Our team has clear "rules" for decision making.
15. Decisions are practically planned for and key information to support decisions is transparently shared.
16. Our team uses the available experience in the organization to shape decision making.

Team Dynamics

17. Our team has an accountable leader who ensures we have what is necessary to do our job.
18. Our team has the necessary experience and expertise to achieve its objectives.
19. Our team recognizes, encourages, and employs team members' individual strengths and preferences.
20. I feel comfortable speaking up and sharing my opinions.
21. Our team has good interpersonal relationships.
22. Working together energizes and uplifts the team.
23. Our team ensures all team members are able to make a full contribution to achieving our goals.
24. We are united and open across functional boundaries.

25. We view problems that arise as a shared responsibility.
26. Our team handles conflict well within the team.

Organizational Support

27. Management provides clear and timely direction when needed.
28. The organization's governance structure allows us to be nimble and punchy.
29. The organization recognizes and reinforces good work done by our team.
30. Our team balances achieving results and considering personal needs [e.g., work/life balance].

Open Questions

31. What is going well for the team?
32. What can be improved to help the team work better together?

Appendix II

List of Common Agile Practices

- Time boxing
- Retrospective
- Spike Solution
- Planning Poker
- Product backlog
- Backlog grooming
- Backlog prioritization
- Progressive elaboration
- Minimum marketable features
- Minimum Viable Product (MVP)
- Minimum Business Increment (MBI)
- Personas
- Story Mapping
- Sprints
- Epics
- User stories
- Visualize workflow
- Wireframe
- Daily stand-up
- Exploratory testing
- Definition of Done
- Theory of Constraint
- Mindfulness
- Avoid waste
- Short iterations
- Sprint goals
- Servant Leader
- Self-organization
- Team agreements
- Release goals
- Release plan
- Project chartering
- Quality Assurance
- Refactoring
- Relative sizing
- Product vision
- Increment
- Exploration

- Limit Work in Progress
- Fast feedback
- Last responsible moment
- Pair programming
- Face-to-face communication
- Osmotic communication
- Burn up/down
- Unit testing
- Technical debt
- Task board
- Swarming
- Regression test
- Test Driven Development
- Just in Time
- Acceptance criteria
- Velocity

Appendix III

New PM Onboarding Checklist

A supportive PMO provides onboarding training to new PMs to ensure consistent understanding and application of services. After a project manager is hired, but before he/she is ready to serve in an organization, he/she needs to be prepared for the specifics inside the organization. A typical readiness checklist should be established in the first 30 days of employment, and looks like this:

- The PM is trained to the organization's structure, quality system, drug and/or device development processes, and the project management process.
- The PM is equipped with the tools to perform the job, including access to the company's authoritative project scheduling system and access to pertinent company project information.
- The PM's project assignments are clear, and if appropriate, the transition plan from the outgoing PM is in place.
- The PM is briefed on the specifics of the project team, their members, assignments, schedules, project plan(s), currently mitigated and unmitigated risks, as well as the budget along with actuals.
- The PM is introduced to the team and the role explained.

Use this checklist to help ensure that new PMs joining your team are set up with the best possible chance of success. Some of these items will be completed prior to the new PM joining; some will continue for several months beyond the start date. The checklist is broken down into different areas for the different roles involved:

- **The formal authority**: PMO, PM's manager, or similar
- **The mentor**: A more experienced PM assigned to support the new PM and help them acclimate to the new environment
- **The PM**: The person who is being brought on board

The best chance of success comes from having a well-rounded approach to onboarding the PM, so it is important to have all three roles involved. Any processes around new employee setup (computer hardware, desk, corporate policies and benefits, etc.) are usually handled as part of existing HR onboarding processes, so they are not included here.

1 Functional Manager: <name of manager>

Item	Date Completed	Notes
Project(s) assigned		
Mentor appointed		
First-day schedule developed and resources confirmed		
Project briefing conducted (purpose, status, work, staff, etc.)		
Weekly checkpoint meetings scheduled		
Sponsor overview meeting scheduled		
Sponsor overview meeting completed		
PMO meeting scheduled		
PMO meeting completed		
Team meeting scheduled		
Team meeting completed		
Process/methodology training scheduled		
Process/methodology training completed		
First-week review scheduled		
First-week review completed		
Second-week review scheduled		
Second-week review completed		
First-month review scheduled		
First-month review completed		
Three-month review scheduled		
Three-month review completed		
Personal goals developed and entered into HR system		
Professional development plan developed		

2 Mentor: <name of mentor>

Item	Date Completed	Notes
Background/résumé reviewed		
Appointed projects reviewed		
Introductory meeting scheduled		
Introductory meeting completed		
Weekly checkpoint meetings scheduled		
Development/concern areas identified		
Skills/knowledge development plans developed		
Skills/knowledge development checkpoints established		
Development accountability model developed		

3 PM: <name of PM>

Item	Date Completed	Notes
All team members met		
All stakeholders met		
Communication plan for all team members/stakeholders developed		
Project assessment completed (challenges, knowledge gaps, etc.)		
Methodology reviewed and questions/uncertainties identified		
Personal development plan developed for all gaps (not formal organizational development)		
Regular meetings with mentor occurring		
Regular meetings with PMO occurring		
Regular meetings with functional manager occurring		
Informal conversations with colleagues/other PMs occurring		
Personal goals agreed to		
Professional development plan agreed to		

Appendix IV
Transition Plan

Name

Job Title

Date of Transfer

Core Activities

Activity	Description	Transfer to

Standing Meetings

Meeting	Description	Transfer to
	• Purpose • Attendees • Meeting time/frequency • Link to meeting materials	

Ongoing Projects/Tasks

Project/Task	Description	Transfer to
	• What is the project/task? • Who are the key team members? • What are the known risks/caveats? • What is unknown? • Where is project information stored?	

Upcoming Projects/Tasks

Project/Task	Description	Transfer to

Appendix V
Resource Management Definitions

Term	Definition
Capacity	The volume of work an organization can perform
Demand	The amount of project work that the organization would like to do
Capacity management	The process of adjusting the number of resources to meet a given demand
Demand management	The process of adjusting the number of projects to meet a given capacity

Appendix VI
Generic Job Description for a Director-Level Development Project Manager

1 Job Summary

The DPM partners with the Development Program Lead (DPL) to manage the Development Project Team (DPT) to successfully shape and execute the project and drug development strategies. The DPM is accountable for the cross-functional execution of one or more integrated Asset Development Plans, including management of timelines, cost, quality, and risk mitigation. The DPM provides an independent voice to shape the project strategy and drive optimal decisions for the broader portfolio value. This position can be assigned to any therapeutic area and work on projects at any stage of development.

2 Roles & Responsibilities

- The DPM leads project management for one or more DPTs focused on complex and high-priority programs. In this capacity, they may provide oversight to other PMs.
- Partners with the DPL to foster a high-performing team and monitor the health and operating efficiency of the team as a unit.
- Leads creation of team norms and operating principles with a focus on cross-functional input and accountability and robust analysis of vetted options.
- Acts as an integrator within the DPT and across the enterprise to ensure alignment and connect best practices.
- Facilitates effective, science-based business decisions, including the development of scenarios as needed. Ensures all decisions are assessed as to their impacts and communicated to stakeholders in a transparent and timely manner.
- Adds strategic value by deriving insights, and has the ability to influence projects against those insights.
- Highlights interdependencies and downstream impacts of strategic decisions.
- Proactively identifies risks and ensures response plans (mitigations and contingencies) are implemented.
- Develops and coordinates resource planning across functions to assure adequate resources are applied to the project.

- Advises on governance expectations with focus on cross-functional input and rigorous debate.
- For projects that are being jointly developed, works closely with a key strategic alliance partner.
- Acts as a Change Agent for continuous improvement and transformational initiatives.
- Coaches project team members and others to drive excellence and accountability and develops talent pipeline for DPT membership.
- May coach and/or mentor more junior PMs and/or PM talent pipeline via rotations and other types of engagements.
- Acts a leader in the PM organization to help build cross-portfolio capability by sharing best practices, connecting across the portfolio, and coaching more junior PMs.

PM Department Leader

- As business conditions require, Director-level PMs may additionally serve as a PM Department Leader, with direct reporting responsibility for three to five individual contributor PMs
- Key expectations of the PM Department Leader are to
  - Provide day-to-day mentoring and coaching of direct reports on PM core competencies and behaviors
  - Ensure new tools and processes are embedded effectively by their direct reports
  - Serve as the manager for direct reports for career development, performance management, and all HR matters

3 Requirements

- Bachelor's degree in Life Sciences, Physical Sciences, or relevant discipline, advanced degree preferred
- Five (5) or more years of demonstrated drug development project management experience with 7+ years of relevant experience in the pharmaceutical industry (e.g., Drug Discovery, Clinical Operations, Technical Operations/CMC, Regulatory Affairs)
- Breadth of drug development expertise with solid understanding of project strategy, interdependencies, and disease content to contribute to strategic discussions
- The ideal candidate for Late Development will have experience with operational study start-up into Registrational/Phase 3 trials with a deep understanding of the necessary steps required for market application submissions in multiple regions (US, EU, Japan, China).
- Broad range of leadership skills, including situational leadership and ability to influence without authority
- Demonstrated ability to collaborate in a cross-functional environment

- Strong communicator able to integrate and succinctly summarize the various parts of a project and effectively tailor messages to audience, including senior leaders
- Ability to resolve complex problems and manage difficult stakeholder situations
- Ability to lead the development of critical path analyses and support scenario planning to achieve goals/timelines
- Excellent Project Management Skills – drives execution while balancing speed, quality, and cost.

Index